A SHORTER HISTORY
OF
SCIENCE

Plate I ISAAC NEWTON

A SHORTER HISTORY
OF
SCIENCE

BY

SIR WILLIAM CECIL DAMPIER
(formerly WHETHAM)
Sc.D., F.R.S.

Fellow and sometime Senior Tutor of Trinity College, Cambridge
Fellow of Winchester College

CAMBRIDGE
AT THE UNIVERSITY PRESS
1944

CAMBRIDGE
UNIVERSITY PRESS

University Printing House, Cambridge CB2 8BS, United Kingdom

Published in the United States of America by Cambridge University Press, New York

Cambridge University Press is part of the University of Cambridge.

It furthers the University's mission by disseminating knowledge in the pursuit of education, learning and research at the highest international levels of excellence.

www.cambridge.org
Information on this title: www.cambridge.org/9781107672901

© Cambridge University Press 1944

This publication is in copyright. Subject to statutory exception
and to the provisions of relevant collective licensing agreements,
no reproduction of any part may take place without the written
permission of Cambridge University Press.

First published 1944
First paperback edition 2014

A catalogue record for this publication is available from the British Library

ISBN 978-1-107-67290-1 Paperback

Cambridge University Press has no responsibility for the persistence or accuracy of URLs for external or third-party internet websites referred to in this publication, and does not guarantee that any content on such websites is, or will remain, accurate or appropriate.

CONTENTS

Preface *page* ix

Chapter I. THE ORIGINS 1
Prehistory—The Dawn of History—Egypt—Babylonia—China and India—Crete and Mycenae.

Chapter II. GREECE AND ROME 16
Greek Religion—The Ionian Philosophers—The Atomists—The Pythagoreans—Greek Medicine—Socrates and Plato—Aristotle—The Hellenistic Period—Deductive Geometry—Archimedes, Aristarchus and Hipparchus—The School of Alexandria—Alchemy—The Roman Age.

Chapter III. THE MIDDLE AGES 34
The End of Ancient Learning—The Fathers of the Church—The Dark Ages—The Arabian School—The Revival in Europe—The Thirteenth Century and Scholasticism.

Chapter IV. THE RENAISSANCE 46
The Causes—Leonardo da Vinci—The Reformation—Copernicus and Astronomy—Chemistry and Medicine—Anatomy, Physiology and Botany—Magnetism and Electricity—Philosophy—Witchcraft.

Chapter V. GALILEO AND NEWTON: THE FIRST PHYSICAL SYNTHESIS 61
Galileo—Boyle, Huygens and others—Scientific Academies—Newton and Gravitation—Mass and Weight—Optics and Light—Newton and Philosophy—Newton in London.

Chapter VI. THE EIGHTEENTH CENTURY 78
Philosophy—Mathematics and Astronomy—Chemistry—Physiology, Zoology and Botany—Geography and Geology—Machinery.

Chapter VII. PHYSICS AND CHEMISTRY OF THE NINETEENTH CENTURY 92
The Scientific Age—Heat and Energy—Thermodynamics—The Wave Theory of Light—Spectrum Analysis—Electric Currents—Electro-magnetism—The Atomic Theory—Organic Chemistry—Chemical Action—Solution—Colloids.

Chapter VIII. NINETEENTH-CENTURY BIOLOGY 113
Biology and its Effects—Physiology—Bacteriology—The Carbon and Nitrogen Cycles—Geography and Geology—Evolution and Natural Selection—Anthropology—Nineteenth-Century Science and Philosophy—Psychology.

Chapter IX. RECENT BIOLOGY 129

Genetics—Geology and Oceanography—Biochemistry, Physiology and Psychology—Viruses and Immunity—Anthropology.

Chapter X. THE NEW PHYSICS AND CHEMISTRY 142

The Physical Revolution—Cathode Rays and Electrons—Radio-activity—X-rays and Atomic Numbers—Positive Rays and Isotopes—The Structure of the Atom—The Transmutation of Elements—Electro-magnetic Waves—Relativity.

Chapter XI. THE STELLAR UNIVERSE 162

The Solar System—The Stars and Nebulae—The Structure of Stars—Stellar Evolution—The Beginning and End of the Universe—Conclusion.

A Note on Bibliography 174

Index 176

PLATES

I.	Isaac Newton	*Frontispiece*
II.	Archimedes	*facing p.* 28
III.	Leonardo da Vinci	29
IV.	Galileo	62
V.	Galileo's original telescope	63
VI.	Charles Robert Darwin	122
VII.	Lord Rutherford of Nelson	123
VIII.	Tracks of α particles in oxygen	148
IX.	The Spiral Nebula in Canes Venatici	149

Plate I is from a photograph by Ramsey and Muspratt; Plate II is from Cox, *Mechanics*; Plate III, Anderson photograph, supplied by Mansell; Plate IV, photograph supplied by Rischgitz; Plate V is from Dampier, *Cambridge Readings in Science*; Plates VI and VII by permission of Elliott and Fry; Plate VIII is from Rutherford, *The Newer Alchemy*; Plate IX is from Jeans, *The Universe Around Us*.

FIGURES IN THE TEXT

1.	Palaeolithic flint tool	*page* 2
2.	Cave drawing of a mammoth	4
3.	Palstave of late middle bronze age, found at Hilfield, Dorset	6
4.	Early Egyptian pottery	9
5.	An Egyptian queen driving in her chariot	10
6.	The Theorem of Pythagoras	19
7.	Drawing of a dissection of development of *Sepia* by Aristotle	23
8.	Diagram of the Universe by Copernicus	51
9.	Diagram of declination of iron magnet	57
10.	Co-ordinate geometry by René Descartes	59
11.	Curve of error	96
12.	Asymmetric carbon atom	107
13.	Curve of error in biology	124
14.	Thomson's apparatus for cathode rays	144

PREFACE

The speed with which three editions of my larger book—*A History of Science and its Relations with Philosophy and Religion*—have been called for shows that many men are interested in the subjects with which it deals. Some, however, have found the philosophic part difficult to read, and have asked for a straightforward story of the growth of science reduced to its simplest terms.

It is impossible to ignore altogether the connexion of science with other activities, but description can, if we will, be confined to the more direct impacts. The Greek Atomists, besides speculating on the structure of matter, developed therefrom a mechanical theory of life. Conversely, the philosophy of Plato, as modified by Aristotle, laid too much stress on innate ideas and logical deduction to make a favourable background for the beginnings of inductive experimental science. To Newton and his immediate followers the Heavens declared the Glory of God, but Newton's work produced a very different effect on the minds of Voltaire and other eighteenth-century sceptics. Darwin's revival of the old theory of evolution on the new basis of natural selection not only suggested an alternative origin for mankind, but spread evolutionary doctrine far beyond the limits of biology. The recent revolution in physics has shaken the evidence for philosophic determinism which the older synthesis seemed to require. Such broad effects must be noticed, but the more technical aspects of philosophy may be passed by.

In writing this 'Shorter History', I have had two objects: firstly, to help the general reader, who wishes to know how science, which now affects his life so profoundly, reached its present predominance, and secondly, to meet the needs of schools. Those older schoolboys whose chief subjects are scientific should look at them also from a humanist standpoint, and realize their setting in other modes of thought, while those studying literature need some knowledge of science before they can be said to be well educated. For both groups, I believe that the history of science, the story of man's attempts to understand the

PREFACE

mysterious world in which he finds himself, makes the best way of approach to common ground. Moreover, an opinion is growing that early specialization is dangerous, an opinion which leads to a desire to give even scholarship candidates a well-balanced education. I hope that this little book may be useful to schoolmasters who share my views, and lead many readers to study its longer prototype.

I wish to thank those who have allowed me to use their illustrations, and the Cambridge University Press for making publication possible in war-time. I must also thank my sister, Miss Dampier, for helping in the tedious work of preparing the index.

W. C. D.

December 1943

A SHORTER HISTORY OF SCIENCE

CHAPTER I. *THE ORIGINS*

Prehistory. What is Science? The word comes from the Latin *scire*, to learn, to know, and thus should cover the whole of learning or knowledge. But by English custom it is used in a narrower sense to denote an ordered knowledge of nature, excluding such humanistic studies as language, economics and political history.

Science has two streams corresponding to two sources, the first a gradual invention of tools and implements whereby men earn their living more safely and easily, and the second the beliefs they form to explain the wonderful universe around them. The first may perhaps better be called technology, for its problems are too difficult for early theory, and only in later stages does it become applied science; the second, which in historic times grew into a pure search for knowledge, is the main subject of this book.

If we seek for the beginning of science and the matrix in which it arose, we must trace the records of early man as given us by geologists and anthropologists, who study respectively the structure and history of the earth and the physical and social characters of mankind.

It is probable that the crust of the earth solidified some thousand million years ago, 1·6 thousand million, or $1·6 \times 10^9$ years, is a recent estimate. Geologists recognize six great periods which followed that event: (1) Archaean, the age of igneous rocks formed from molten matter; (2) Primary or Palaeozoic, when life first appeared; (3) Secondary or Mesozoic; (4) Tertiary; (5) Quaternary; (6) Recent. The age of these periods relative to each other is shown by the position of their deposits in the earth's strata, but no certain estimate can be made of their absolute age in years.

One school of anthropologists holds that traces of man's handiwork are first seen in tertiary deposits, and the most recent evidence is held to support this view. The earliest signs of man, perhaps somewhere between one and ten million years ago, a minute fraction of the earth's life, are flints or other hard stones roughly chipped

THE ORIGINS

into tools or weapons. They are found lying on the surface of the earth, in river beds, in excavations made by engineers or dug deliberately to find them, and in caves—one of the most primitive types of dwelling. The oldest stone tools, named eoliths, are difficult to distinguish from natural products, flints chipped accidentally by the action of water or movements in the earth, but the next group or palaeoliths are clearly artificial and of human origin (Fig. 1).

Ignoring the doubtful eoliths, we can divide the stone age into two parts. Palaeolithic man only chipped his implements; he hunted wild animals, but did not tame them or cultivate the soil. Neolithic man belonged to a different and higher race, which seems to have invaded Western Europe, bringing with it domestic animals, some skill in agriculture and in the forming of pottery and of polished implements in flint or hard igneous stone, in bone, horn or ivory. In some parts of the world Neolithic man found out how to smelt copper and harden it with tin, thus passing from the stone to the bronze age, and incidentally making the first discovery in metallurgy. Later on, bronze gave place to iron, probably because of its greater advantage in weapons of war.

Fig. 1. Palaeolithic flint tool

Returning to a consideration of the stone age, we see that the variety and finish of the tools found increase as we examine the higher, and therefore later, deposits. Weapons dropped in war or the chase give us only casual finds, but occasionally we come upon the floor or hearth of a prehistoric dwelling, and add more largely to our collection. Signs of fire, such as burnt flints, show another agent in the hands of man, while the remains of plants and animals indicate by their nature the climate of the time, whether warm, temperate or glacial.

At an early stage in the story, man took to living in caves, a shelter from the weather ready to his hand, and a museum kept for us, containing not only dropped tools and weapons, but also, beginning in Palaeolithic times, pictures drawn on the walls by the inhabitants,

THE ORIGINS

pictures from which we can gain some knowledge of the life lived by men thousands or millions of years ago, and even an insight into their thoughts and beliefs.

Lower Palaeolithic civilizations, dating from the beginning of the quaternary era, and ending as the last ice age approached, must have covered an immense stretch of time, during which there seems to have been a steady improvement in culture, at all events in the lands which are now England and France.

Middle Palaeolithic times are associated with what is known as Mousterian civilization, so named from the place where it was first discovered—Moustier near Les Eyzies. The race which made it, known as Neanderthal man, again from its place of discovery, was of a low type, generally held not to be in the direct line of human evolution. The cold of Mousterian times drove man more extensively to caves and rock shelters as homes, and so preserved many of his tools, which show that he had learnt to fashion them from flakes chipped off flints, unlike the majority of Lower Palaeolithic tools, which were made from the cores left when flakes were chipped away.

Upper Palaeolithic or Neo-anthropic man appeared in France after the worst of the last ice age was over, though a continuing mixture of reindeer with stag in the bones and pictures shows that the climate was still cold. The Upper Palaeolithic race was far higher in the scale of humanity than any earlier one, and began to make household objects; there was a definite bone industry and the flaking of flint was greatly improved. We can see such things as eyed needles and double-barbed harpoons carved in bone, and found in Magdalenian, Upper Palaeolithic, deposits. These and other tools and weapons show a marked advance on earlier implements.

Of somewhat the same age are the oldest pictures on the walls of caves. Outlines of men and animals—horses, buffaloes and extinct mammoths—appear. Then, as some indication of beliefs, we have drawings thought to represent devils and sorcerers.

To gain a more definite idea of these beliefs, we may compare them with those of early historical times, as described by Greek and Latin authors, and those still found among primitive people in various parts of our modern world. A huge amount of such evidence has

4   *THE ORIGINS*

been collected by Sir James Frazer in his great book *The Golden Bough*, primarily to explain the rites of Diana Nemorensis, Diana of the Wood, carried on, even in classical days, in the Grove of Nemi in the Alban Hills near Rome, and obviously surviving from earlier, more barbarous, ages. In the Grove of Nemi lived a priest-king who reigned there until a man, stronger or more cunning than he, slew him and held the kingship in his stead.

FIG. 2. Cave drawing of a mammoth

In order to explain this tragic custom, Frazer ranges over the world and over long stretches of time. He deals in turn with magic, magical control of nature, nature spirits and gods, human gods, gods of vegetation and fertility, the corn-mother, human sacrifices for the crops, magicians as kings, the periodic killing of kings, especially when crops fail or other catastrophes happen, and the arts as an approach to primitive science. Some anthropologists regard magic as leading directly to religion on one side and to science on the other, but Frazer thinks that magic, religion and science form a sequence in that order. Another anthropologist, Rivers, holds that magic and primitive religion arise together from the vague sense of awe and mystery with which the savage looks at the world.

THE ORIGINS

Magic assumes that there are rules in nature, rules which, by the right acts, can be used by man to control nature. Thus magic is a spurious system of natural law. Imitative magic rests on the belief that like always produces like. When frogs croak, it rains. The savage feels he can do that too; so, in a drought, he dresses as a frog and croaks to bring the wished-for rain. Countless similar instances of imitation might be given. Contagious magic believes that things once in contact have a permanent sympathetic connexion. The possession of a piece of another man's clothing, and still more of a part of his body—his hair or his nails—puts him in your power; if you burn his hair, he too will shrivel up.

Now these examples take the magician no farther; by coincidence his action may sometimes be followed by the appropriate happening, but more often it fails. Suppose, however, that, by accident, the magician hits on a real relation of cause and effect—for illustration he rubs together two bits of wood and produces the miracle of fire. By that experiment he has learnt a true fact which he can repeat at will, and he has, for that one relation, become a man of science. But in magic, if he fails too often to produce his effect, he may be forsaken or even killed by his disappointed followers, who may perhaps cease to believe in the control of nature by men and turn to propitiate imagined and incalculable spirits of the wild, gods or demons, in order to obtain what they want; thereby they pass to some form of primitive religion. Meanwhile, far out on the other wing, the discovery of fire, the taming of animals, the growing of crops, the gradual improvement in tools, and the development of many other simple arts, lead, by a less romantic but surer road, to another origin of science. Whatever may be the relation between magic, religion and science—and that relation may differ in various times and places—there is certainly a real and intimate connexion between them. Science did not germinate and grow on an open and healthy prairie of ignorance, but in a noisome jungle of magic and superstition, which again and again choked the seedlings of knowledge.

Neolithic man again shows an advance. Structures such as Stonehenge, where a pointer stone marks the position of the rising sun at the solstice, serve not only religious uses but astronomical functions also.

THE ORIGINS

Prehistoric burials often give interesting information. They are found till the end of Neolithic times, cremation only appearing commonly in the bronze age, and then mostly in Central Europe where forests supplied fuel. Well-finished stone implements were often placed in the tombs, showing us the state of contemporary art, and sometimes suggesting a belief that such things would be useful to the dead when they passed over to another world—a belief then in survival.

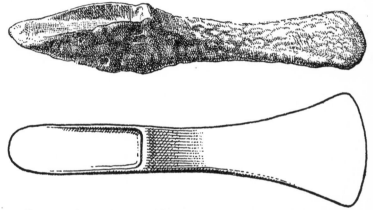

Fig. 3. Palstave of late middle bronze age, found at Hilfield, Dorset

We must not assign to primitive people, whether prehistoric, ancient or modern, too rationalist an outlook. When a savage dreams of his dead father, he does not reason about it, but accepts the dream as real, and his father as, in some sort, alive—not perhaps as much alive as his mother still in this world, but quite clearly surviving his death, though perhaps in an attenuated form as a spirit or ghost. There is no distinction in kind between natural and supernatural, only a vague difference in degree.

With the coming of the bronze age, we pass to a higher culture, made possible by the use of metal. We find axes, daggers and their derivatives spears and swords, and household goods such as lamps. Man has definitely passed from the use of stone as his sole material,

THE ORIGINS

and when bronze is replaced by iron, we approach and soon enter periods when true history can be pieced together by written records on stone, clay, parchment or papyrus.

The Dawn of History. Settled life, with primitive agriculture and industrial arts, seems first to have begun in the basins of great rivers—the Nile, the Euphrates with the Tigris, and the Indus, while analogy would suggest that the early civilization of China too began near its rivers. In contact here and there with these river folk were nomads, pastoral people, wandering with their flocks and herds over grass-clad steppes or deserts with occasional oases. Nomad society was, and is, essentially patriarchal, the social unit being the family, perhaps with slaves, and the government the rule of the father. In normal times, the units kept separate from each other, each in search of food for their beasts. In the Old Testament we have an early and vivid account of the life of nomad people.

And Lot also, which went with Abram, had flocks, and herds, and tents. And the land was not able to bear them, that they might dwell together.... And Abram said unto Lot... Is not the whole land before thee? separate thyself, I pray thee, from me: if thou wilt take the left hand, then I will go to the right; or if thou depart to the right hand, then I will go to the left.[1]

With these isolationist views and customs, neither civilization nor science was possible. Co-operation between the family groups only arose for some definite purpose—a hunt of dangerous wild beasts, or war with other tribes. But sometimes, owing to a prolonged drought, or even perhaps to a permanent change in climate, the grass failed, the steppes or oases in the deserts became uninhabitable, and the nomad folk overflowed as an irresistible horde, flooding the lands of the settled peoples as barbarous conquerors. We can trace several such outrushes of Semites from Arabia, of Assyrians from the borders of Persia and of dwellers in the open grass-clad steppes of Asia and Europe.

Now it is clear that we need not look among nomads for much advance in the arts, still less for the beginnings of applied science.

---

[1] Genesis xiii. 5-9.

But the Old Testament, preserved as the sacred books of the Jews, not only gives in its earlier chapters an account of nomads, but later on deals with the legends of the settled kingdoms in the Near and Middle East—Egypt, Syria, Babylonia and Assyria—a good introduction to the more recent knowledge obtained by the discovery of the buildings, sculptures and tablets, and by the excavations of such relics as royal tombs. This later knowledge is of course fragmentary, depending upon the double chance of the survival of ancient records, and of their discovery and correct interpretation by present-day explorers.

Since the late stone age, the sea-coasts and islands of the Aegean have been chiefly occupied by the Mediterranean race, short in stature with long-shaped head and dark in colouring; to them is due what prehistoric advance in civilization occurred. Farther inland, especially among the mountains, the chief inhabitants were and are of the so-called Alpine race, a stocky people of medium height and colouring, and broad, round-shaped skulls, who pushed into Europe from the east. Thirdly, spreading out from the shores of the Baltic, we find a race which may be called Nordic, tall, fair-haired and, like the Mediterranean people, with long-shaped heads.

Egypt. Egypt is divided into two very different parts—the Delta, where the Nile seeks the sea through mud flats of great fertility, and the valley of Upper Egypt, spreading for a few miles broad in the rift through which the river makes its way between the sands of the western desert and the rocky hills of the eastern shore.

On both sides of the Nile rift, and at many points along its length, Palaeolithic implements are found, showing that human occupation began in early geological times. Then came Neolithic man, with his better tools and the great discovery of the potter's art. When soft clay is worked into a designed form and fixed in that form by fire, a new thing has been created. Here is something more than adaptation, as in flint weapons or stone bowls, something which means true invention and a long step towards civilization.

The history of Egypt begins with the first dynasty of kings, somewhere about 3000 years before Christ. Earlier times are repre-

THE ORIGINS

sented by legends such as that of the divine Horus, Sun of the Day, and his followers. The hieroglyphic script, in which many Egyptian records are written, was first deciphered from the Rosetta stone, discovered by Boussard in 1799 near Rosetta, east of Alexandria. On it is set out a decree of Ptolemy V Epiphanes, in hieroglyphics, demotic script[1] and Greek. Thus those hieroglyphics were interpreted by Champollion and Young, and a beginning made in the study of Egyptian documents, both on incised stone and on paper made from papyrus which, in the dry climate of Egypt, does not perish.

The first clearly historical king was Menes (3188–3141 B.C.), who became sovereign of all Egypt and founded the city of Memphis. Even under his, the first, dynasty, records were kept of the chief events of each year, such as the height of the Nile flood. Documents become plentiful in the time of the fourth dynasty, and the pyramids, orientated astronomically, were built from the fourth to the twelfth dynasty, but the best achievements in practical arts begin to appear under the eighteenth dynasty, somewhere about 1500 to 1350 B.C.

FIG. 4. Early Egyptian pottery

The Egyptians imagined a divine intervention to explain the origin of every craft, art or science; especially were they referred to Thoth, a moon god who measured time in days and years, and established in the temples 'watchers of the night' to record astronomical events.

In arithmetic a decimal system was employed as early as the first dynasty. It had no sign for zero (a much later Indian invention) and no positional notation, also no separate signs for numbers between 1 and 10, so that the unit had to be repeated to the number required. Fractions also were dealt with by units, so that our $\frac{7}{12}$ was written as $\frac{1}{3}$ $\frac{1}{4}$ implying addition. These unit fractions persisted long after mixed fractions became general.

[1] A shortened form of hieroglyphics.

Fig. 5. An Egyptian queen driving in her chariot

The official calendar contained 365 days in the year, though the Egyptians seem to have known that the solar year was nearer $365\frac{1}{4}$. The former year thus worked back through the seasons, completing a cycle in about 1500 years, a period which appears to have been taken to mark an era of time. Five days were held to celebrate the birthdays of five chief gods, and were not counted in the temples, which kept a year of 360 days. This calendar of course diverged even more rapidly from the true year. By the time of the Romans, the many calendars in use brought confusion, but Julius Caesar, with the technical advice of Sosigenes, accepted and established as the Julian calendar a year of $365\frac{1}{4}$ days—365 with an additional day every fourth year. This, being a little too much, was corrected by Pope Gregory XIII in 1582, and in England in 1752, by dropping out one leap year in three centuries out of five.

The periodic submersion of the ground by the Nile, with the consequent loss of boundary-marks, led to the art of land measurement by surveyors or 'rope stretchers', an art which later, in the hands of the Greeks, became the science of deductive geometry.

The stars were grouped in constellations, which were identified with deities, and so represented on ceilings and coffin lids. The Universe was imagined as a rectangular box, with Egypt at the middle of its base. The sky was supported by four mountain peaks, and the stars were lamps hung from the sky by cables. Round the land ran a river, on which travelled a boat bearing the sun.

The Egyptians had a considerable amount of medical knowledge, and several known papyri contain notes or treatises on medicine mingled with magic. The first physician whose name has survived is I-am-hotep, 'he who cometh in peace'; after death he became a god of medicine. The custom of embalming the dead led naturally to anatomy and so to surgery, illustrated in carvings as early as 2500 B.C. The number of diseases treated rationally gradually increased, but mental disorders continued to be referred to exorcists, who, with amulet and charm, professed to drive out the evil spirits.

Ancient Egypt made great advances in technical arts. The stupendous building work of the pyramids is still a wonder, and the ruins of temples at Karnak and Philae show high artistic merit. Set squares, levels and plumb lines are among the tools found. The beam balance was used for weighing, and many weights have been discovered, but it seems that different standards were used in different parts of the country. A weaving loom is pictured as early as the twelfth century B.C., so that spinning must also have been practised. A ship under full sail is pictured in a tomb of the fourth dynasty; also wheeled chariots drawn by horses (see p. 10).

To sum up—the science of ancient Egypt was the handmaid of practical arts, housekeeping, industry, architecture, medicine, but in that rôle it achieved considerable success. Egypt had much influence on other lands, especially on Crete in the Minoan ages and afterwards on Greece, where its practical science was sublimated into a pure search for knowledge.

Cambyses, son of Cyrus the Persian, conquered and crushed ancient Egypt. The thirtieth dynasty, 400 B.C., was the last native line, and another barbarian, Ochus, completed the ruin. When Alexander entered the country, he was hailed as a deliverer; he founded Alexandria, which later on succeeded Athens as the intellectual centre of the world. In 30 B.C., after the reign of the great Cleopatra, Egypt became a Roman province.

Babylonia. Babylonia is the country of the rivers Tigris and Euphrates, and Babylon was for long the huge capital city, of which only ruins remain. Other cities were numerous, one of the oldest being Ur of the Chaldees, from which Abraham set forth on his wanderings.

The country consisted of two parts, the upper, a land of steppe and desert, and the lower, formed by the silt and mud of the rivers, the fertility of which made the wealth of the people. On the west was Arabia, and on the east Assyria, on the foothills of the Persian mountains.

The first name of the country was Sumer, and its earliest inhabitants in historical times were called Sumerians. These people were afterwards mixed with Semites invading from Arabia, and Assyrians from the hills. The land is subject to violent storms and floods, the memory of one such being preserved in the legend of Noah and the deluge. Like this wild nature with its dangers, the Babylonian gods were mostly inimical to man.

The Babylonian Universe was, like the Egyptian, a box, the Earth being its floor. In the centre were snow mountains in which was the source of the Euphrates. Round the Earth was a moat of water, and beyond it celestial mountains supporting the dome of the sky. More useful than such speculations was astronomical observation, which can be traced back to about 2000 B.C. by records found on clay tablets. By the sixth century, the relative positions of the Sun and Moon were calculated in advance and eclipses predicted.

On the basis of such definite knowledge, a fantastic scheme of astrology was built up, in the belief that the stars controlled human affairs. Chaldaean astrologers were specially famous, while sorcerers and exorcists acted as physicians.

Wheat and barley seem to have been indigenous, and were cultivated for food at an early date. This made a calendar necessary; the day as a unit of time is forced on man, and attempts to measure the number of months in the cycle of the seasons were made at a date said to be about 4000 B.C. The day was divided into hours, minutes and seconds, and a vertical rod or gnomon set up as a sundial. Seven days were named after the Sun, the Moon and the five known planets, and thus the week became another unit of time.

But perhaps the most striking scientific achievement was the recognition of the need for fixed units of measurement, and the issue on royal authority about 2500 B.C. of standards of length, weight and capacity. The unit of length for instance was the finger, about $\frac{2}{3}$ inch; the cubit contained 30 fingers, the surveyor's cord 120 cubits and the league 180 cords or 6·65 miles.

The Sumerians too had some skill in mathematics and engineering. They had multiplication tables and lists of squares and cubes. Decimal and duodecimal systems were used, special importance being assigned to the number 60 as a combination of the two, as seen in the measures of minutes and hours. Simple figures and formulae for land surveying led to the beginning of geometry. Maps of fields, towns, and even of the then known world, were drawn. But all were mixed with magical conceptions, which later passed westward. European thought was dominated for centuries by Babylonian ideas of the virtues of special numbers and the prediction of the future by geometrical diagrams.

China and India. The civilization of China approaches in age that of Egypt or Babylonia, but China was isolated, and had no early contact with those countries. There are legends ranging back to some such time as 2700 B.C., but historical records only appear about 2000. Fine pottery and bronze vessels are found in the age of the Shang dynasty, 1750 to 1125, and iron weapons were first used about 500 B.C., later than in Europe. By 100 B.C. trade had been established with Persia and other countries. Culture reached a high level under the Chou dynasty (say 1100 to 250), with practical arts like agriculture and the irrigation of land.

The earliest religion, Taoism, was inextricably mixed with magic. Confucius in the sixth century before Christ introduced a purer faith, and established an Academy, where the literary classes were separated from the priestly. The teaching of Buddha reached China in A.D. 64. In later times China has the credit of an independent invention of paper, and the discovery of the magnet with its use in the compass.

Indian Culture seems to have begun in the valley of the Indus, in the third millennium before Christ. The original dark-skinned people were mixed in the north of India by an incursion of Aryan invaders, who impressed their language and civilization on the former inhabitants. A decimal notation was already in use, and, in the third century, the scheme of numerals we employ to-day was invented; though it reached us through the Arabs, and is known erroneously as Arabic. In ethical philosophy, Buddha (560 to 480 B.C.) is pre-eminent, and schools of medicine with famous physicians had been established in his time.

A primitive atomic theory was formulated or borrowed from the Greeks, and, a century or so before Christ, the idea of discontinuity was extended to time. Everything exists but for a moment, and in the next moment is replaced by a facsimile; a body is but a series of instantaneous existences—time is atomic.

Crete and Mycenae. The early civilization of the Eastern Mediterranean is best known to us by the researches begun by Sir Arthur Evans at Cnossus in Crete, the birthplace and chief home

THE ORIGINS

of what is now called Minoan Culture, from Minos, the legendary king of Crete. Its beginnings, Early Minoan, seem to be contemporary with dynasties I to VI in Egypt; Middle Minoan with dynasty XII, while Late Minoan corresponds to dynasty XVIII, and can be dated 1600 to 1400 B.C. There seems to have been constant intercourse in trade and otherwise between the two countries.

The Cretan script is like, but not identical with, that of Egypt; its early pictorial characters had come to have phonetic meanings when we first find them; they are still undeciphered, but we have much information from other sources. There are remains of skilfully engineered roads crossing the mountain passes; applied science in the palace—mechanical, hydraulic and sanitary; mural decoration with pictures showing costumes (some worthy of modern Paris), weapons and armour, and light chariots drawn by horses.

The Palace at Cnossus was destroyed by civil disturbance or foreign enemies about 1400 B.C. It is worth noting that, at somewhat the same time, Egypt was attacked by sea-borne raiders, who resemble in many ways the Achaean invaders of Greece.

But we can look to the mainland for another site of Eastern Mediterranean culture. Rich relics of a civilization similar to that of Crete have been found at and near Mycenae. But about 1400 the Aegean lands were in turmoil and Mycenae like Cnossus went down.

The earliest of Greek traditions speaks of the coming of the Achaeans, apparently brown-haired tall men from the grass-lands lying to the north, immediately perhaps from the valley of the Danube, bringing with them horses and iron weapons to conquer the bronze of the Minoans, and the custom of burning instead of burying the dead. It is thought by some that here we see the first incursion of the Nordic race—tall, fair, and with long-shaped skulls—filtering through the pastures and steppes of Europe from their original home by the shores of the Baltic. However that may be, they were followed about 1100 by the Dorians, also probably from the nearer north, who overran the Peloponnese. By this mixture of peoples a nation was created; Minoan civilization gave place to Greek, and we enter the full light of history.

CHAPTER II. *GREECE AND ROME*

Greek Religion. As Xenophanes recognized as long ago as the sixth century before Christ, whether or no God made man in His own image, it is certain that man makes gods in his. The gods of Greek mythology first appear in the writings of Homer and Hesiod, and, from the character and actions of these picturesque and for the most part friendly beings, we get some idea of the men who made them and brought them to Greece. The men differ in customs and beliefs from the Minoans and Myceneans, and it is probable that they were the Achaeans, who came down from somewhere in the north.

But ritual is more fundamental than mythology, and the study of Greek ritual during recent years has shown that, beneath the belief or scepticism with which the Olympians were regarded, lay an older magic, with traditional rites for the promotion of fertility by the celebration of the annual cycle of life and death, and the propitiation of unfriendly ghosts, gods or demons. Some such survivals were doubtless widespread, and, prolonged into classical times, probably made the substance of the Eleusinian and Orphic mysteries. Against this dark and dangerous background arose the Olympic mythology on the one hand and early philosophy and science on the other.

In classical times the need of a creed higher than the Olympian was felt, and Aeschylus, Sophocles and Plato finally evolved from the pleasant but crude polytheism the idea of a single, supreme and righteous Zeus. But the decay of Olympus led to a revival of old and the invasion of new magic cults among the people, while some philosophers were looking to a vision of the uniformity of nature under divine and universal law.

The Ionian Philosophers. The first European school of thought to assume that the Universe is natural and explicable by rational inquiry was that of the Ionian nature-philosophers of Asia Minor. One of the earliest known to us is Thales of Miletus (*c.* 580 B.C.), merchant, statesman, engineer, mathematician and astronomer. Thales is said to have visited Egypt, and, from the empirical rules for land-surveying there in vogue, to have originated the science of deductive geometry. He pictured the Earth as a flat disc, floating on water, instead of resting on a limitless solid bottom, and propounded the idea of a cycle from air, earth and water through the bodies of plants and animals to air, earth and water again.

Anaximander (*ob.* 545) recognized that the heavens revolve round the pole star, and inferred that the visible dome of the sky is half a complete sphere, with the Earth at its centre, the Sun passing underground at night. Worlds arise from the primordial stuff of chaos by natural causes, such as are still at work. The first animals came from sea slime, and men from the bellies of fish. Primary matter is eternal, but all things made from it are doomed to destruction.

Some men, such as Empedocles of Sicily, held that there were four elements, earth, water, air and—still more tenuous—fire. By combinations of the four, the various types of matter were made. Empedocles proved the corporeal nature of air by showing that water can only enter a vessel as air escapes.

The Ionian philosophy was brought to Athens by Anaxagoras of Smyrna about 460 B.C. Anaxagoras added to its mechanical bent by the belief that the heavenly bodies are of the same nature as the Earth; the Sun being not the god Helios but a burning stone.

The Ionians also made advances in practical arts, inventing or importing the potter's wheel, the level, the lathe, the set square and the style or gnomon, used as a sundial to tell the time and to determine when the Sun's altitude at noon was greatest.

The Atomists. The intellectual heirs of the Ionians were the Atomists—Leucippus, who founded a school at Abdera in Thrace some time in the fifth century, and Democritus, born there in 460 B.C. Their views are known to us by references in later authors such as Aristotle, by the work of Epicurus (341 to 270), who brought the atomic theory to Athens, and by the poem of the Roman Lucretius two centuries later.

The Atomists taught that everything happens by a cause and of necessity. They carried farther the Ionian attempt to explain matter in terms of simpler elements. Their atoms are identical in substance but many in size and shape; 'strong in solid singleness', they have existed and will exist for ever. Thus difference in the properties of bodies is due to difference in size, shape and movement of atoms. The appearances as seen by the senses have no reality: 'according to convention there is a sweet and a bitter, a hot and a cold, and according to convention there is colour. In truth there are atoms and a void.'

Literary men sometimes claim for the Greeks anticipations of modern science. But the modern atomic theory which Dalton sponsored in the early years of the nineteenth century was based on the definite facts of chemical combination in equivalent weights, and was soon verified experimentally by predictions obtained by its aid, as were the later developments of the theory due to J. J. Thomson, Aston, Rutherford and Bohr. In the absence of such clear experimental evidence, the atomic theory of Democritus and Lucretius is little more than a lucky guess. If many different hypotheses are put forward, one of them may well chance to be somewhere near more modern views. The atoms of the Greeks had no firm basis, and were upset by the equally baseless criticism of Aristotle. The Greek atomic theory was more philosophy than science.

Nevertheless the men of Ionia and the Atomists were closer to a scientific attitude of mind than were some other schools in ancient times. They, at all events, tried to explain the world on rational lines. They failed; but they had the credit of what, in another connexion, is now called 'a near miss'.

The Pythagoreans. As a contrast to the rational and materialist philosophy of the Ionians and Atomists, we find systems of thought formulated, perhaps direct from Orphism, by those of more mystical minds. One of the earliest and greatest of such men was Pythagoras, who was born at Samos, but moved to Southern Italy about the year 530 B.C. In spite of his mysticism, he was a mathematician and an experimenter, and, on the latter account, was accused by Heraclitus of practising 'evil arts'.

The Pythagoreans were the first to emphasize the abstract idea of number, irrespective of the actual bodies counted. In practice this made arithmetic possible, and in philosophy it led to the belief that number lay at the base of the real world, though it was difficult to reconcile with another Pythagorean discovery—the existence of incommeasurable quantities. The idea of the importance of number was strengthened when experiments with sound showed that the lengths of strings which gave a note, its fifth and its octave were in the simple ratios of $6:4:3$. The distance of the planets from the earth must conform to a musical progression, and ring forth 'the music of the spheres'. Since $10 = 1+2+3+4$, ten was the perfect number, and the moving luminaries of the heavens must be ten also; as only nine could be seen, there must be somewhere an invisible 'counter-earth'. Aristotle rightly blamed this juggling with facts.

But the Pythagoreans recognized the Earth as a sphere, and saw that the apparent rotation of the heavens could best be explained by a revolving Earth, though they thought that, balanced by the counter-earth, it swung round, not the Sun, but a fixed point in space—thus they did not fully anticipate Aristarchus and Copernicus as is sometimes said.

Pythagoras and his followers carried farther the science of deductive geometry—indeed the forty-seventh proposition of the first book of Euclid is known as the Theorem of Pythagoras. Though the equivalent 'rule of the cord' for laying out a right angle may have been long known in Egypt, it is probable that Pythagoras was the first to prove by deduction from axioms that the square on the hypotenuse of a right-angled triangle is equal to the sum of the squares on the other two sides.

The mystic view was also seen in Heraclitus (c. 502), poet and philosopher, to whom the primary element was a soul-stuff of which the world is made, though all things are in a state of flux—πάντα ῥεῖ. Again, Alcmaeon the physician regarded man the microcosm as a miniature

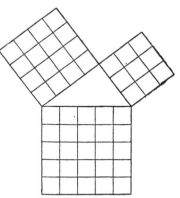

FIG. 6. The Theorem of Pythagoras

of the Universe or macrocosm; his body reflects the structure of the world, and his soul is a harmony of number. These fanciful ideas reappear both in classical times and in the Middle Ages.

But the Pythagorean doctrines were opposed by other contemporary philosophers. Zeno of Elea thought he had discredited the theory of numbers and also multiplicity by a series of paradoxes. Achilles pursuing a tortoise reaches the spot whence the tortoise started; but the tortoise has now moved on to a further place; when Achilles gets there it has again moved; Achilles can never catch the tortoise. The paradoxes rest on misconceptions about the nature of infinitesimals, of time, and of space, only to be cleared up in the nineteenth century with the recognition of different kinds of infinity.

Greek Medicine. It is probable that Greek medicine owed much to Egypt. There were two schools—those of Cos and Cnidos, the former placing most trust in *vis medicatrix naturae*. In our own day it has been said that, while the difference between a good doctor and a bad one is immense, the difference between a good doctor and none at all is very small. Perhaps Cos had made the same discovery. Cnidos, having more (or less) faith, searched for a specific remedy for each disease.

Greek medicine reached its zenith in Hippocrates (460–375), called 'the Father of Medicine', with a theory and practice of the art in advance of the ideas of any period till modern times. Observation and experiment appeared, with conclusions based on inductive reasoning, while many diseases were accurately described and fairly appropriate treatment indicated.

Socrates and Plato. A considerable advance in knowledge was due to the historians, who described the nature of countries as well as their histories. The greatest were Hecateus (540–475), Herodotus (484–425) and Thucydides (460–400). The latter gave an eye-witness's account of the Peloponnesian war and described the plague at Athens and the solar eclipse of the year 431.

A reaction from atomism held that since sensation certainly exists, while its messages about reality are doubtful, sensation itself is the only reality. And so we come to Socrates (470–399), whose work we know chiefly from the writings of Xenophon and Plato. Socrates was primarily an educationalist, using the dialectic method of question and answer to eliminate opinions he thought false and suggest those he thought true, exposing with inimitable humour ignorance, stupidity and pretentiousness wherever he found them. He rejected mechanical determinism and indeed all the conflicting theories of the physicists, and held that their search for a knowledge of reality was a futile attempt to transcend the limits of human intelligence. He regarded the mind as the only worthy subject of study; the true self was not the body, but the soul and inward life. Thus in a sense he led a religious reaction, though popular clamour charged him with newfangled atheism. But Aristotle credits Socrates with two scientific achievements—universal definitions and inductive reasoning.

Socrates' disciple Plato (428–348) deduced theories of nature from human needs and desires, and showed the influence of the Pythagorean mystical doctrines of form and number. His astronomy was crude, though he regarded the stars as floating free in space, moved by their own divine souls. He initiated the idea of cycles, to represent the apparent path of the Sun round the Earth—an idea afterwards developed in detail by Hipparchus and Ptolemy.

Plato's cosmos was a living organism with body, soul and reason, Alcmaeon's macroscopic analogy of man the microcosm. On such ideas Plato's science was, for the most part, fantastic. Moreover, he despised experiment as a base mechanic art or blamed it as impious, though he prized highly the deductive subjects such as mathematics. But he invented negative numbers, and treated the line as flowing from a point—the basic idea of Newton's method of fluxions.

Thus Plato was led to his famous theory of 'intelligible forms'— the doctrine that ideas or 'forms' alone possess full reality, the doctrine that, in later ages, became known as 'realism'. When the mind begins to frame and reason about classes and definitions, it finds itself dealing not with individuals but with these hypothetical types or forms. Natural objects are in a constant state of change, it is only the types or forms that are constant and therefore real. There is a clear analogy between the ideas or forms of Plato and the abstractions we now find inherently necessary for science, abstractions which enter into our formulation of scientific concepts and our reasoning about their relations. Plato was very near some kinds of modern philosophy, but his idealism was not calculated to help experimental science at that stage of the world's history.

Aristotle. Aristotle (384–322), born at Stagira in Chalcidice, was a son of the physician to Philip, king of Macedon, and was himself the tutor of Alexander the Great. He was the most notable collector and systematizer of knowledge in the ancient world, and, till the Renaissance in modern Europe, no one approached him in scientific learning. The early Middle Ages had only incomplete compendiums of his writings, and their attitude of mind was chiefly Platonic; but in the thirteenth century Aristotle's full works were recovered, and finally made the basis of the philosophy of Scholasticism by Saint Thomas Aquinas, a philosophy which was only replaced, after four hundred years of dominance, by the science of Galileo and Newton. Hence comes Aristotle's supreme importance in the history of thought.

Aristotle was at his best as a naturalist. Biology till lately was merely an observational science, giving little opportunity for the speculative theories so dear and so dangerous to the Greeks. Aristotle mentions five hundred animals, some with diagrams gained by dissection. He described the development of the embryo chicken, detected the formation of the heart, and watched it beat while yet in the egg. In classification he saw that it was well to use as many distinguishing qualities as possible to bring together animals nearly related.

Even in physiology Aristotle made advances, insisting on the need for dissection. He considered the structure and function of lungs and gills, and, having little knowledge of chemistry to help him, concluded that the object of respiration was to cool the blood by contact with air—a false idea, but perhaps the best possible at the time. The brain too was to Aristotle a cooling organ, the seat of the intelligence being in the heart, though Alcmaeon and Hippocrates had already placed it in the brain.

Aristotle rejected the atomic theory. Democritus taught that in a vacuum heavier atoms would fall faster than light ones; Aristotle, with an equal absence of evidence, argued that in a vacuum all must fall equally fast, but, as this conclusion is inconceivable, there can never be a vacuum, and the theory of atoms, which requires it, is false. If all things are made of a common material, all would be heavy, and nothing tend to rise in seeking its 'natural place'. No one till Archimedes understood what we now call density or specific gravity —the weight per unit volume—on which, compared with that of the

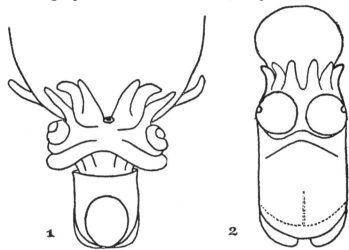

Fig. 7. Drawing of a dissection of development of *Sepia* by Aristotle

surrounding medium, rise or fall depends. Unlike the objective physics of Democritus, who sought to explain nature by material atoms, the concepts of Aristotle were attempts to express man's perceptions of the world in terms natural to his mind in those days, categories such as substance, essence, matter, form, quantity, quality, concepts acceptable to the Greeks and Mediaevalists, but to us vague and unsatisfying. Aristotle accepted the spherical form of the Earth, but still regarded it as the centre of the Universe.

Aristotle was a disciple of Plato, but, being himself chiefly engaged in the study of individual animals and other bodies, he did not follow Plato into extreme idealistic 'realism'. He held in a modified form what afterwards came to be called 'nominalism', which gives reality to individuals, though Aristotle allowed some reality also to 'forms' or universals. Thus Aristotle was a philosopher as well as a naturalist. Moreover, he created formal logic with its syllogisms and show of complete proofs. It was a great achievement, and could be applied well to deductive subjects like mathematics, especially geometry, as we shall see later. But syllogistic logic has little to do with experimental science, where inductive discovery and not deduction from accepted premises is the first object sought; deduction only comes in when, a hypothesis having been framed, we wish to predict consequences fit for experimental examination. Aristotle's influence did much to turn Greek and mediaeval science into a hopeless search for absolutely certain premises, and a premature use of deductive methods. Some think that here we have one of the chief reasons why little advance was made in science for two thousand years. Nevertheless, Aristotle himself regarded theories as temporary expedients; it was in later ages that his views were made fixed and rigid.

The Hellenistic Period. With Alexander of Macedon we enter a new epoch. As he marched to the East, he took with him Greek learning, and in turn brought Babylonia and Egypt into closer touch with Europe, while his staff, in the first co-operative researches, collected facts about geography and natural history. Gradually the nations were linked together, the upper classes by Greek culture, and the rest by a universal Greek dialect, ἡ κοινή, the common speech, understood 'from Marseilles to India, from the Caspian to the Cataracts'. Commerce became international and thought free.

The general philosophic systems of Athens gave place to more modern methods, definite and limited problems were specified and attacked singly, and real progress in natural knowledge was made. The change somewhat resembles the overthrow of mediaeval scholasticism by the clear-cut science of Galileo and Newton. Though other places were involved, the centre of this new learning was Alexandria, the Egyptian city founded in 332 B.C. by Alexander the Great.

The Greek element was predominant, but Babylonian astronomy became available in Greek translations, bringing with it the fallacies of Chaldean astrology. The fantastic idea of the relation between the cosmos as macrocosm and man as microcosm strengthened the belief in the control of man by the planets in their courses, and in a heedless Fate behind them which rules stars and gods and men. To meet this terrible Babylonian creed, men sought any means of escape from Fate. They looked firstly to the heavens themselves, where apparently incalculable bodies like comets offered some hope of freedom; secondly to magic, which promised control of nature, the papyri of the time being full of recipes for charms and spells. But, absurd as it seems, we must not forget that, in some parts of England and Wales, a belief in magic, charms and spells lies a very little way beneath the surface even to-day.

At the same time the Mystery Religions also spread from the East. These religions, based on prehistoric rites of initiation and communion, sought salvation by personal union with a Saviour God, known under many names, who had died and risen again. With the breakdown of the Olympians and the local deities, in the second century before Christ, and onwards till the rise of Christianity, men's deepening religious sense was mostly met by the Mystery Religions.

Astrology, magic and religion make their appeal to the many, philosophy and science to the few. The most important Hellenistic philosophy was Stoicism, taught in Athens about 315 B.C. by Zeno of Citium. It spread both east and west, till it became the chief philosophy of Rome. Its theology was a form of pantheism with a high and stern concept of morality.

The Stoic had little to do with science, which is more concerned with the revival of atomism by Epicurus of Samos (342–270), whose writings preserved its tenets till Lucretius recorded them two centuries later. Epicurus led a reaction against the idealist philosophy of Plato, holding that all is corporeal and death is the end of life. Gods exist, but they are products of nature and not its creators and are 'careless of mankind'. But man is free, subject neither to capricious gods nor to blind remorseless fate. The only test of reality is man's sensation. To Epicurus as to Democritus, all things are made of atoms and void, and our world is but one of many, formed by the chance conjunctions of atoms in infinite space and endless time.

Thus Epicurus built a system of cheerful if shallow optimism on the atomic theory and a primitive sensationalism.

Deductive Geometry. Perhaps the most successful and characteristic product of the Greek mind was deductive geometry. Started by Thales about 580 B.C. on Egyptian land-surveying, it was developed by Pythagoras, approved by Plato and Aristotle, and the existing knowledge collected and systematized by Euclid of Alexandria about 300 B.C. in such a complete way that his work was the standard text-book of geometry in my own youth. From a few axioms regarded as self-evident, a series of propositions was deduced by logical principles, the model of a deductive science.

But nowadays we look at deductive geometry in one of two possible ways, both different from that of Euclid. Firstly, it can be taken as the intermediate, deductive step in a science observational and experimental. The axioms and postulates are really hypotheses as to the nature of space, reached by induction from the observed phenomena; the observer deduces what he can from his hypotheses, and this process forms the science of deductive geometry. But to complete the whole it is necessary to see whether or no the consequences agree with observation or experiment, all on the general lines of an inductive, experimental science. Till quite lately, astronomy verified Euclid; it required Einstein's alternative hypothesis about space, and the power of modern telescopes, to show that Euclidean space, though in accordance with facts to a high degree of accuracy, was not the whole truth; it breaks down when tested to the utmost limit.

Secondly, we can look at the problem in another way. Observation suggests space of a certain kind, and from this the mind defines an ideal space which *is* exactly what actual space seems to be. We can then, with no reference to nature, develop the logical consequences of the definition without asking if they correspond with observation. If, for instance, space is defined as an extension of three dimensions, one set of consequences—those of Euclid—follows. If we start afresh and define what corresponds to space as having n dimensions, other results follow. It is a pretty exercise in pure mathematics, but it has nothing necessarily to do with nature or experimental science.

But of course both these methods are essentially modern; the Greeks accepted the simple belief that space is what it seems to be, and the axioms of Euclid's geometry self-evident facts. But nevertheless Greek geometry was a great achievement, a step in mathematical science which never had to be retraced.

Archimedes, Aristarchus and Hipparchus. And now we come to one of the greatest—perhaps the greatest—name in Greek science, Archimedes of Syracuse (287–212), who laid the foundations of mechanics and hydrostatics (see Plate II, facing p. 28). Certainly he is the Greek with the most modern scientific outlook. The origins of these sciences are to be found in the practical arts, and this primitive knowledge Archimedes put into form by a combination of experiment and the deductive methods learned in geometry: hypotheses are set forth, their consequences deduced and then compared with the results of observation and experiment.

The use of the lever is illustrated in the sculptures of Assyria and Egypt two thousand years before Archimedes, but, with the Greek love of deduction, Archimedes deduced the law of the lever from axioms which he regarded as self-evident: (1) that equal weights placed at equal distances from the point of support balance, and (2) that with equal weights at unequal distances the one at the greater distance descends. Implicitly the law of the lever, and its equivalent the principle of the centre of gravity, are contained in these axioms, but it is well to have the connexion formally proved. Nowadays we treat the law of the lever as a matter of experiment, and deduce other mechanical results from this basis: it is just a question of where to begin.

The idea of density which, unknown to Aristotle, caused him to go so far astray, became clear to Archimedes. The story is that King Hiero, having given gold to artificers to make him a crown, suspected them of alloying it with silver and asked Archimedes to test his suspicion. Archimedes, thinking over the problem in his bath, noticed that his body displaced an equal volume of water, and saw in a flash that for equal weights the lighter silver alloy would displace a larger volume of water than the heavier gold. Exclaiming '$εὕρηκα$,

εὕρηκα', Archimedes is said to have leapt from his bath, presumably to test the crown and then tell Hiero of his discovery. When a body floats in a liquid, its weight is equal to the weight of liquid displaced, and, when it is immersed, its weight is diminished by that amount. This is called 'the principle of Archimedes'.

But Archimedes' chief interest lay in geometry, and he thought more of his geometrical discoveries than of the practical results—compound pulleys, hydraulic screws, burning mirrors—for which he was famous among the people; he thought them the recreations of a geometer at play. The modernity of his outlook caused his works to be eagerly sought for at the Renaissance, especially by Leonardo da Vinci. His writings were nearly lost to the world; at one time the only survival was one manuscript of the ninth or tenth century A.D. which disappeared. But three copies had been made, and from them the text was recovered. Archimedes was killed at the storming of Syracuse in the year 212. His tomb was found and restored in 75 B.C. by Cicero, then Quaestor in Sicily.

Meanwhile an older contemporary of Archimedes, Aristarchus of Samos (310–230), had made a great advance in astronomy in his still existing work *On the Sizes and Distances of the Sun and Moon*. By applying some capable geometry to eclipses of the Moon and to its appearances when half full, Aristarchus concluded that the ratio of the diameter of the Sun to that of the Earth must be greater than 19 : 3 and less than 43 : 6, that is about 7 : 1. This figure is of course much too small, but nevertheless the method is sound, and the result that the Sun is larger than the Earth was in itself a revolutionary achievement. But both Archimedes and Plutarch say that Aristarchus went much farther, and advanced the theory that the fixed stars and the Sun remain unmoved, while the Earth revolves round the Sun in a circular orbit. Sir Thomas Heath calls Aristarchus 'the Ancient Copernicus'.

But Aristarchus was too much in advance of his age, though, according to Plutarch, Seleucus the Babylonian held the heliocentric view in the second century B.C. and sought new evidence to support it. For the rest, the belief in the solid Earth beneath man's feet as the

Plate II ARCHIMEDES

Plate III   LEONARDO DA VINCI. Self Portrait at Turin

centre of the Universe was too strong even for philosophers and astronomers.

Hipparchus was born at Nicaea, and worked at Rhodes and then at Alexandria from 160 to 127 B.C. He used the older Greek and Babylonian records, invented many astronomical instruments and made accurate observations therewith, dividing his circles into 360 degrees in the Babylonian manner. He discovered the precession of the equinoxes, measured this slow motion and the distance of the Moon, and got them nearly right. He was a competent mathematician, and invented both plane and spherical trigonometry.

His cosmogony was wrong in its fundamental assumption, but the idea that the Sun, Moon and planets were carried round the Earth in crystal spheres or cycles, while on these were superposed smaller orbits or epicycles, though complicated, could be made to represent the facts, and, expounded by Ptolemy of Alexandria about A.D. 127 to 151, held the field till the sixteenth century. And, as long as the Earth was the centre, the follies of astrology almost inevitably arose again and again. Plato had heard of astrology, but it was first really brought to Athens from Babylon by Berossus about 280 B.C., and recurred from time to time through the ages. One would have thought that it would have been effectively put down by Copernicus, Galileo and Newton, but, I am told, a belief in astrology is still prevalent among the uneducated in all classes.

The School of Alexandria. One of Alexander's generals, Ptolemy (not to be confused with the astronomer), founded in Alexandria a Greek dynasty which lasted till the death of Cleopatra in 30 B.C. The reign of the first Ptolemy (323–285) was made illustrious by Euclid the geometer and Herophilus the anatomist and physician.

About the middle of the third century before Christ the famous Museum was founded at Alexandria. It had departments of literature, mathematics, astronomy and medicine, which were research institutes as well as educational schools. They all used the greatest library of ancient times, containing some 400,000 volumes or rolls, and reckoned one of the wonders of the world. But one section of the library was destroyed by the Christian Bishop Theophilus about A.D. 390, as a stronghold of pagan learning, and the rest, whether accidentally or wilfully, by the Muhammedans in 640—one of the greatest losses which the human mind has suffered. The library must have contained copies of the works of countless authors now for ever gone, some of their very names perhaps perished.

In medicine the leading Alexandrians were Herophilus, Erasistratus and Eudemus. Herophilus, living in the reign of Ptolemy I, was the earliest great human anatomist, and the most famous physician since Hippocrates. He described the brain, nerves and eye, the internal organs, arteries and veins, and proved that the site of intelligence is the brain and not the heart as taught by Aristotle. Erasistratus, a younger contemporary of Herophilus, also practised human dissection and made experiments on animals, treating physiology for the first time as a separate subject. He held the atomic theory and its accompanying philosophy in opposition to medical mysticism.

Meanwhile there had been increases in geographical knowledge, beginning in the fourth century before Christ. Hanno, passing the Pillars of Hercules at the western exit from the Mediterranean, sailed down the west coast of Africa; Pytheas voyaged round Britain; and Alexander marched to India. It came to be generally accepted that the Earth was a sphere, and some idea of its size was formed. The variation with latitude in the length of day and night led Ecphantus to the idea of the revolution of the Earth on its axis, though it was still placed at the centre of space.

The first great physical geographer was Eratosthenes, Librarian of the Museum at Alexandria. By measuring the distance apart and the latitudes of two places on the same meridian, he calculated the circumference of the Earth as 24,000 miles, a result surprisingly close to the modern estimate of 24,800. He conjectured that the Western Ocean might be divided by a New World, though later Poseidonius, underestimating the Earth's size, proclaimed that an explorer sailing west for 70,000 stades (7000 or 8000 miles) would come to India. This it was that inspired Columbus. Ptolemy too was a geographer as well as an astronomer, and made maps of the then known world.

GREECE AND ROME

Apollonius the mathematician carried geometry into the study of various curves, and Hero the mechanician invented contrivances such as the siphon, a thermoscope and a primitive steam-engine.

Alchemy. In early times industries arose to supply imitations of jewels and other things too expensive for the people, for instance, cheap alloys which looked like silver and gold. Matter itself, Plato and other philosophers believed, was unimportant, but its qualities were real. The mordant salts used in dyeing will etch metals, and, if a small quantity of gold has been added, usually to an alloy of tin, lead, copper and iron, whitened by the addition of mercury, arsenic or antimony, etching may leave a golden surface; the gold, they believed, acting as a ferment, changed the whole into the nobler metal. Such facts and such ideas were linked with others, and particularly with astrology. The Sun generates his image, gold, in the body of the Earth, the Moon represents silver, the planet Venus copper, Mercury quicksilver, Mars iron, Jupiter tin and Saturn the heavy and dull metal lead. The gods had been moved from Olympus to the sky, but there, as Sun, Moon and planets, they continued to control the destinies of man, and help him to obtain the noble metals his soul longs for. Qualities clearly can be changed, and qualities are the essence of things—indeed philosophically *are* the things. On such an underlying basis of practice and beliefs, the alchemists, allied with the astrologers, were started and encouraged in their vain quest for an elixir of life to cure all ills, and a philosopher's stone to transmute base metals into gold. Their search failed, but, in its course, the astrologers helped astronomy, and the alchemists made discoveries on which was founded the science of chemistry. Astrology came originally from Babylon, but alchemy made its first clear appearance in Alexandria during the first century of the Christian era, and lasted there for three hundred years. The Alexandrian alchemist was neither a fool nor a charlatan, but an experimenter, acting in accordance with the best philosophy of his age; it was the philosophy which was at fault. When alchemy revived among the Arabs and in Europe, the prevalent philosophy and terminology both had altered, and the new alchemists, not understanding the Alexandrians, tried to change the substances themselves, and not merely their qualities. They usually hid their failures in a flood of mystical verbiage.

The Roman Age. The Romans had great efficiency as lawyers, soldiers, and administrators, but little creative intellectual power; the art, science and medicine of Rome were borrowed from the Greeks. The Romans only cared for science as a help in practical life. They ignored the essential basis, knowledge sought for its own sake, and in a few generations pure science, and following it applied science, ceased to advance.

The chief Roman philosophy was Stoicism, founded by Zeno of Citium, brought to Rome by Diogenes the Babylonian, and best known in the writings of Seneca and of the Emperor Marcus Aurelius. Stoicism acquired a tincture of Platonism from Poseidonius, more famous as a traveller, and geographer, who explained the tides by the joint action of the Sun and Moon.

The atomism of the Greeks was expounded by Lucretius (98–55 B.C.) in his poem *de Rerum Natura*. The poem aims at the overthrow of superstition by the acceptance of the atomic and mechanical philosophy, causation controlling all things, from the invisible vapour of water to the mighty skies, the flaming walls of the world—*flammantia moenia mundi*.

Among the writers of compendiums were Pomponius Mela a geographer, and Cicero the statesman (106–43 B.C.), who wrote *de Natura Deorum*, which contains some information about the scientific knowledge of the time. Both Virgil and Varro wrote on agriculture, and 100 years later the elder Pliny (A.D. 23–79) produced in his *Naturalis Historia* a much more complete encyclopaedia of the whole science of his period—geography, man and his qualities, animals, plants and trees, together with agriculture, fruit-growing, forestry, wine-making, the nature and use of metals, and many of the fine arts. He described the lion, the unicorn and the phoenix, the practice and special utility of various forms of magic, all with equal acceptance. Plutarch (A.D. 50–125) wrote on the nature of the Moon and on Roman mythology, with a suggestion for the comparative study of religions. Both he and Diogenes Laertius handed on valuable information about Greek philosophy and philosophers.

Greek medicine flourished at both Alexandria and Rome. Celsus in the reign of Tiberius wrote a treatise on medicine and surgery, describing many surprisingly modern operations. He gives also a history of medicine in Alexandria and Rome. His work was lost in the Middle Ages, but recovered in time to influence the Renaissance. Later on Galen (A.D. 129–200) dissected animals and a few human bodies, and investigated by vivisection the action of the heart and the spinal cord. His medicine was built on the idea of spirits of various kinds pervading the different parts of the body, and from this theory he deduced dogmas which influenced medicine for 1500 years. Galen's πνεῦμα ψυχικόν, translated as *spiritus animalis*, became our 'animal spirits', the original meaning of which is perhaps sometimes misunderstood. Galen thought that the principle of life was a pneuma, drawn from the world spirit by the act of breathing. He held that the blood reached the parts of the body by a tidal ebb and flow through the veins and arteries, a view which persisted till Harvey discovered circulation. In the first century A.D. Dioscorides, a military physician, wrote on botany and pharmacy, describing some six hundred plants and their medicinal properties. The schools of medicine, being of practical use, especially for the armies, lasted longer than pure science and philosophy, which, during the second century A.D., showed signs of the collapse that soon followed.

But practical arts of all kinds flourished in Rome. The protection of public health, with aqueducts to bring fresh water, sanitary appliances, hospitals and a public medical service, military and civil engineering, showed the practical genius of the Roman people.

Diophantus of Alexandria, the greatest Greek writer on algebra, lived in the latter half of the third century after Christ. He introduced abbreviations for quantities and operations that continually recur, and thus was able to solve simple equations. After him there was no one of the first rank in mathematics or pure science in the ancient world. Indeed no advance in knowledge was being made, and the only activity shown was in the writing of compendiums and commentaries, chiefly on Greek philosophers.

And these second-hand sources were all, or nearly all, which came through the dark ages and lightened the dawn which followed.

CHAPTER III. *THE MIDDLE AGES*

The End of Ancient Learning. To understand why for a thousand years Europe made little or no advance in science we must begin by examining the scientific and philosophic outlook and the religious beliefs which dominated thought during the last period when the learning of Athens, Alexandria and Rome was still alive.

As long as his books were available, Aristotle was accepted as the great authority on scientific theory and even fact, and this acceptance remained when, in the sixth century, his complete works ceased to be read or were lost, and only abbreviated commentaries were studied, the most popular being one on Logic.

In spite of Aristotle's predominance in the little science that survived, philosophy took another course. Plato's School at the Academy in Athens was by then teaching a mystical Neo-Platonism, and Platonism in its mystical Neo-Platonic form became the prevalent philosophy. As we shall see presently, this Platonism survived as an alternative background even in the later Middle Ages of Aristotelian Scholasticism, and, at the time we are now considering, the first few centuries of the Christian era, it was predominant.

Alongside the inquiries of the philosophers, classical Greek mythology still represented the official religion, though it showed signs of decay. Even at their height, the many and various pagan cults were tolerant of each other, and the policy of the Roman Empire upheld that tolerance. But the first Christians, taking over the exclusive religious outlook of the Jews, refused to conform to the Imperial laxity, and thus faced their Roman governors with a difficult problem, leading almost inevitably to persecution. But when Christianity began to triumph and persecution of its votaries ceased, the need of a more definite formulation of its beliefs arose.

The Fathers of the Church. Thus the first systematic theology of the Christian Church was framed by the Early Fathers by interpreting the Christian story in the light of Hebrew Scripture and

Neo-Platonic philosophy, while underneath still lay the mystery religions and more primitive magical rites. In these rites we find the widespread ideas of initiation, sacrifice and communion, which appear in more developed forms in the mystery religions, and also find the celebration in imagery of the drama of the year—growth and full life in summer, death in winter, and joyous resurrection in each new spring.

Among the Eastern mystery religions two are of special interest: Mithraism, the Persian cult of the soldiers' god Mithras, disputed for long with Christianity the possession of the Roman Empire; Manichaeism, which held a dualism of the powers of good and evil, reappeared in essence again and again as the ages passed. .

Such was the atmosphere in which the Early Fathers sought, as their most open and obvious task, to reconcile Greek philosophy and Christian doctrines, while traces of magic rites were unconsciously interwoven in their synthesis. Foremost among the Fathers was Origen (185–254), who proclaimed the essential conformity of the ancient learning, especially Alexandrian science, with the Christian Faith, and did much to gain converts among the educated and intelligent. Origen took a more critical view of the Old and New Testaments and a more liberal-minded outlook generally than afterwards became orthodox after much embittered controversy. It may be held that doctrines which, for any reason, survived were apt to be accepted as orthodox, while, as Gibbon says, 'the appellation of heretics has always been applied to the less numerous party'.

Of the Latin Fathers, Saint Augustine (354–430) had most influence on Christian thought. He was successively a Manichaean, a Neo-Platonist and a Christian, and combined Platonic philosophy with the teaching of Saint Paul's Epistles to form the first great Christian synthesis of knowledge. It was this combination of Neo-Platonism and Pauline Christianity which persisted in the background through the predominance of Aristotle when his works were recovered in the late Middle Ages.

Early Christian theology and mystic Neo-Platonism acted and reacted, each accusing the other of plagiarism. At first little stress was laid on divination and magic, but a century later the Neo-

Platonists Porphyry and Iamblichus and the Christians Jerome and Gregory of Tours revelled in the daemoniac and miraculous.

Again, the personal motive of individual sin, salvation or damnation, and the prospect of a catastrophic end of the world at the near second coming of Christ, replaced the bright Greek spirit and the stern Roman joy in Family and State.

With such an outlook on life and death, it is no wonder that the Christian Fathers showed small interest in secular knowledge for its own sake. As Saint Ambrose said, 'To discuss the nature and position of the earth does not help us in our hope of the life to come'; and the followers of Artemon are said 'to lose sight of heaven while they are employed in measuring the earth'. Thus part of the Christian world passed to an actual hatred of pagan learning. A branch of the Library of Alexandria was destroyed by Bishop Theophilus about the year 390, and Hypatia, the last mathematician of Alexandria, was murdered in 415 with revolting cruelty by a Christian mob, instigated, it was believed, by the Patriarch Cyril.

The Dark Ages. The fall of Rome was not only or even primarily an overthrow of civilization by barbarians. It was more the clearing away of a doomed and crumbling ruin. The military strength of the Empire was weakening; the purity of its blood was being corrupted by an influx of oriental and other foreign elements; a failure of the gold and silver mines of Spain and Greece reduced the currency and caused a fall in prices, which, as in our own day, sent land out of cultivation. The Roman Campagna consequently became infested with malaria, and large tracts grew uninhabitable. The barbarians came, and Rome fell into the dark gloom of the sixth and seventh centuries.
. Almost the only survival was medicine, monastic and other. In the sixth century the Benedictines began to read compendiums of the works of Hippocrates and Galen, which were soon afterwards studied at Salerno, a city on the Bay of Paestum to the south of Naples, the first secular home of learning. Salerno was a Greek colony and later a Roman health resort, and it is possible that it remained an unbroken link between the learning of the ancient and that of the modern world.

Neo-Platonism was given its final form by Proclus (A.D. 411–485), the last great Athenian philosopher, whose works led from the writings of Plato and Aristotle to the mystical beliefs of mediaeval Christianity and Islam.

Again Boëthius, a noble Roman who was put to death as a Christian in 524, wrote compendiums and commentaries on Plato and Aristotle, and on what he called the quadrivium—arithmetic, geometry, music and astronomy. He was the last to show the true spirit of ancient philosophy, and his works form another link between the Classics and Mediaevalists. After Boëthius the classical spirit vanished, and Plato's school at Athens was closed in A.D. 529 by order of the Emperor Justinian.

Curiously enough, countries at a distance from Rome showed some of the earliest signs of revival. In Ireland, Scotland and the north of England, a literary and artistic development, quickened by Christianity, culminated in the work of the Anglo-Saxon monk Bede of Jarrow (673–735), and, together with missionary teaching, carried some secular learning into more southern lands. Bede's science, as that of other mediaevalists, was chiefly drawn from Pliny's *Natural History*, though he added something of his own as, for instance, observations on the tides. Then we come to the Abbey schools founded by Charlemagne, and Alfred the Great's translations of Latin texts into Anglo-Saxon. New nations were forming and becoming self-conscious.

The Arabian School. When European learning was at its lowest ebb, culture of mixed Greek, Roman and Jewish origin survived in the Imperial Court at Byzantium and in the countries which stretch from Syria to the Persian Gulf. The Persian School of Jundishapur gave refuge to Nestorian Christians[1] in A.D. 489 and to Neo-Platonists when Plato's Academy was closed in 529. In the third century, Greek was replaced in Western Asia by the Syriac language, which was itself later replaced by Arabic.

Muhammad and his Arabs between 620 and 650 conquered Arabia, Syria, Persia and Egypt. About 780 Harun-al-Rashid, most

[1] Followers of Nestorius, a sect declared heretical.

famous of the Caliphs, encouraged translations from Greek authors, and thus helped to initiate the great period of Arab learning. The first task of the Arabs was to recover the hidden and forgotten store of Greek knowledge, then to incorporate it in their language, and finally to add their own contributions.

The Arabs developed an atomic theory based on those of the Greeks and Indians. But to the Arabs not only matter, but space and time also were atomic—the latter being composed of indivisible 'nows'. Allah creates atoms anew from moment to moment with their qualities. If He ceased re-creating, the Universe would vanish like a dream.

The Arab schools of medicine and chemistry were developed in the late eighth and in the ninth centuries, when the lead definitely passed from Europe to Arabic-speaking lands. The earliest forms of practical chemistry are concerned with the arts of life—cooking, metal-working, the collection of medicinal plants and the extraction of drugs. The Arabs too worked at alchemy for seven hundred years, searching for the transmutation of metals and an *elixir vitae* to cure all human ills. They were, of course, doomed to failure, but, in their search, they discovered many chemical substances and reactions, on which the true sciences of chemistry and chemical medicine were founded. The results reached Europe chiefly through the Moors in Spain.

The most famous Arabian chemist and alchemist was Abu-Musa-Jabir-ibn-Haiyan, who flourished about 776. He is thought to be the author of many writings which appeared afterwards in Latin and were assigned to a shadowy 'Geber' of uncertain date. Their identity or difference cannot be decided until all Jabir's works have been translated and studied. Jabir seems to have prepared what we now call lead carbonate and separated arsenic and antimony from their sulphides; he described processes for the refinement of metals, the dyeing of cloth and leather and the distillation of vinegar to give concentrated acetic acid. Metals, he believed, differed because of the varying proportions of sulphur and mercury in them. The theory that sulphur (fire), mercury (liquidity) and salt (solidity) were the primary principles of things lasted as an alternative to the atomic theory and the four elements of Empedocles and Aristotle till the days of Robert Boyle's *Sceptical Chymist* published in 1661.

In the ninth century also, translations into Arabic were made of Euclid's *Geometry* and Ptolemy's *Astronomy*, thereafter best known by its Arabic name of Almagest. The Hindu numerals were completed by the invention of a sign for zero, and replaced the clumsy Roman figures in Europe.

Among Muslim astronomers, Muhammad-al-Batani of Antioch recalculated the precession of the equinoxes and a new set of astronomical tables. About the year 1000, observations on solar and lunar eclipses were made at Cairo by Ibn Junis or Yūnus, perhaps the most eminent Muslim astronomer.

The greatest period of Arabian science dates from the tenth century, when al-Kindi wrote on philosophy and physics, and the Persian physician Abu-Bakr-al-Rāzi practised in Baghdad and produced a treatise on measles and small-pox. He was reckoned the leading physician of Islam, a chemist who applied his knowledge to medicine, and a physicist who used the hydrostatic balance. But the most famous Arabic physicist was Ibn-al-Haitham (965–1020), who worked in Egypt chiefly at optics. Translated into Latin, his writings had much influence, especially on and through Roger Bacon. In the eleventh century al-Hazen was famous in mathematics, and Ibn Sīnā or Avicenna (980–1037) wrote on all the sciences then known. His *Canon*, or compendium of medicine, was one of the highest achievements of Arabic culture, and afterwards became the text-book of medical studies in the Universities of Europe. A contemporary was al-Bīrūni, philosopher, geographer and astronomer.

At this time Arabic had become the classical language of learning, and Arabic writings carried the prestige which formerly and afterwards was given to Greek. In the eleventh century Omar Khayyám did important algebraic work, but by the end of that century a decline of Arabic and Muslim learning had set in.

In Spain, where Arabian, Jewish and Christian cultures were intermingled, philosophy developed on much the same lines as in

Christendom a century later. There was the same attempt to reconcile the sacred books of the nation with Greek learning, and the same contest between theologians who relied on reason and those who trusted to revelation or to mystic religious experience, and in either case denied the validity of human reason in matters of faith.

The chief fame of the Spanish-Arabian School is due to Averroes, born at Cordova in 1126, to whom religion was a personal and inward power, and theology an obstacle. His teaching came into conflict with that of orthodox Christian theologians, but, in spite of opposition, he became an authority in the Universities, worthy, it was said, to be compared with Aristotle as a master of the science of proof. Another great Cordovan was the Jewish physician Maimonides (1135–1204), who was also a mathematician, astronomer and philosopher. His chief work was a Jewish Scholasticism, comparable with the Arabic Scholasticism of al-Ghazzali of Baghdad and the Christian Scholasticism of Saint Thomas Aquinas, each of them an attempt to reconcile a particular scheme of theology with the philosophy of Aristotle.

The Revival in Europe. When Arabic learning percolated into Europe, it found here and there the way prepared. The Byzantine Empire at Constantinople patronized art and literature in the ninth and tenth centuries, collecting and preserving many Greek manuscripts. Salerno had developed a school especially of medicine, and the encouragement of scholars by Charlemagne and Alfred had so much improved teaching in the north that the secular schools began to take their modern form of Universities.

Legal studies revived in Bologna about the year 1000, and soon medicine and philosophy were added. A students' Guild or *Universitas* was formed for mutual protection, and undertook also the provision of teachers, so that the governing power rested with the students. But, early in the twelfth century, a school of dialectic was organized by teachers at Paris, and a Community or *Universitas* of teachers set the fashion for most Universities in Northern Europe. Thus at Oxford and Cambridge the governance rests with the teachers, while the election by the undergraduates of the Rectors in Scottish Universities shows a surviving trace of students' control on the model of Bologna.

THE MIDDLE AGES

The academic subjects of study consisted of an elementary trivium—grammar, rhetoric and dialectic, dealing with words—and a more advanced quadrivium—arithmetic, geometry, music and astronomy, supposed to be concerned with things. Music contained a half-mystical doctrine of numbers, geometry a selection of Euclid's propositions without proofs, while arithmetic and astronomy were used chiefly to fix the date of Easter. When later on philosophy also was studied, it was merely added as a more advanced part of dialectic. All led up to the sacred subject of theology; mediaeval philosophy, it has been said, was but a mixture of logic and theology.

The controversy between Plato and Aristotle on the nature of 'universals' or 'intelligible forms' passed into the writings of Porphyry and Boëthius, and so reached the mediaeval mind as the problem of classification. How is it that we can classify? Are individuals alone real as the nominalists say, the classes or universals existing merely as mental concepts, or are the universals apart from the individuals the only realities? Or is Aristotle's compromise to be accepted? Are we to say *universalia ante rem* with Plato, *universalia in re* with Aristotle, or *universalia post rem* with the nominalists? To the Greeks the question was important, and it has some bearing on modern physics. But the mediaeval mind eventually discovered in it the whole problem of Christian dogma. The only difficulty was to determine which alternative was orthodox, and indeed the answer varied from time to time.

At first Plato's realism in a mystical Neo-Platonic form was combined with Christian theology to form the first Mediaeval (as opposed to Patristic) synthesis, based on the idea that ultimately the divine is the only reality. Then nominalism appeared in Berengarius of Tours (999–1088) and in Roscellinus, who reached a Tritheistic conception of the Trinity. In Abelard, on the other hand, the Trinity was reduced to three aspects of one Divine Being. It will be seen that the apparently harmless philosophical problem opens the way to very dangerous controversies.

Abelard showed signs of independence in many directions, saying that 'it is necessary to understand in order to believe', which may well be compared with Anselm's *credo ut intelligam*, and Tertullian's *credo quia impossibile*.

Abelard's desire to understand leads our story to the great enlightenment of the thirteenth century, but before we pass on we must pause to trace the effect on the Middle Ages of the fantastic idea of the cosmos as macrocosm and man as microcosm—an idea we can see in Plato's *Timaeus*, in Alcmaeon's medicine, and perhaps in older legends of Hermes or the Egyptian god Thoth. The mediaeval mind seems to have been fascinated with this analogy. Pictures supposed to represent it are continually seen in mediaeval art—paradise in the empyrean space beyond a zone of fire, and hell in the earth underfoot. The sun, stars and planets, kept moving by the four winds of heaven, are related to the four elements of earth and the four humours of man. Drawings derived from these confused imaginings, I am told, still decorate certain almanacs beloved of the ignorant.

The Thirteenth Century and Scholasticism. In the thirteenth century an increasing desire for secular knowledge led to the rendering of Greek writings into Latin, first by re-translation from the Arabic and later by direct translation from the Greek. The former was carried on chiefly in Spain from 1125 to 1280, and the latter at first in Sicily and Southern Italy, where lived both Arabs and Greeks, who kept up diplomatic and commercial relations with Constantinople. The Arabic nations and the Jews dwelling among them had a real interest in science, and it was by contact with them that mediaeval Europe learned a more rationalist habit of mind.

Between 1200 and 1225 Aristotle's complete works were recovered and translated into Latin, first from Arabic versions and then straight from the Greek. In the latter task, one of the most active scholars was Robert Grosseteste, Chancellor of Oxford and Bishop of Lincoln, while his famous pupil, Roger Bacon, wrote a Greek grammar. Their object was not merely literary; they sought primarily to unlock the original language of Scripture and Aristotle.

The effect of the new knowledge on current controversies was immediate. Realism remained, but it became less Platonic and less dominant. Aristotle's form of realism could be reconciled with nominalism, and made a philosophy on which scientific inquiry could be based. Aristotle's outlook was at once more rational and more scientific than the mystical Neo-Platonism which had come to represent ancient philosophy. His range of knowledge, especially his knowledge of nature, was far greater than anything then available. To absorb and adapt this new material to mediaeval Christian thought was a difficult task. Aristotle's works were condemned by a Church Council at Paris in 1209, but the flood proved irresistible, and in 1225 the University of Paris accepted Aristotle's works as a subject of study.

Two scholars took the foremost place in interpreting Aristotle: Albertus Magnus (1206–1280) and his famous pupil Saint Thomas Aquinas (1225–1274). Albertus combined Aristotelian philosophy with all contemporary knowledge of astronomy, geography, zoology, botany and medicine, and Thomas Aquinas continued his work of rationalizing both sacred and profane learning. Thomas recognized two sources of knowledge: the Christian faith transmitted through Scripture, the Fathers and the tradition of the Church, and the truths reached by human reason as set forth by Plato and Aristotle. The two sources must agree, as both come from God, and a *Summa Theologiae*, as written by Saint Thomas, should comprise the whole of knowledge. On this basis was built up the system of Scholasticism.

Unlike the 'realist' outlook of earlier days, which feared the use of reason by 'nominalists', the Thomist philosophy was founded and built up on reason, and on reason professed to explain the whole of existence, with the object of apprehending both God and nature.

Thomas accepted all Aristotle's logic and science. His logic, based on the syllogism, set out to give rigorous proofs from definite and certain premises, which to Thomas were intuitive axioms and Catholic doctrine: it was ill adapted to guide men in the experimental examination of nature.

With both Aristotle's science and Christian doctrine, Saint Thomas Aquinas took over the assumption that man is the centre and object of creation, and that the cosmos can be defined in terms of human sensation. He also adopted the Ptolemaic system of astronomy with its central Earth, though only as a working hypothesis. This caution, however, was not followed by his disciples; the geocentric theory became part of the orthodox scheme, and afterwards proved a serious obstacle to Copernicus and Galileo. But, accepting Thomas's premises, his whole system, worked out with consummate skill, hung together convincingly, and a criticism of Aristotle's philosophy or science became an attack on the Christian faith.

There was one man of whom thirteenth-century records remain who saw the need of experiment. Roger Bacon, born about 1210 near Ilchester in Somerset, studied at Oxford under Adam Marsh, a mathematician, and Robert Grosseteste the Chancellor. 'In our days', said Bacon, 'Lord Robert, lately Bishop of Lincoln, and Brother Adam Marsh were perfect in all knowledge.' He also names Master John of London and Master Peter de Maharn-Curia, a Picard, as mathematicians, and Peter also as a master of experiment.

Grosseteste was the first to invite Greeks to come from the East as instructors in the ancient form of their language, still read at Constantinople. As said above, Bacon wrote a Greek grammar, and argued that ignorance of the original tongues was the cause of the errors in theology and philosophy of which he accused the Scholastic doctors of the time. He insisted that the only way to verify or disprove their statements was to observe and experiment. He also realized the importance of mathematics, both as an educational exercise and as a tool in experimental science, and this at a time when mathematics and astronomy had earned a bad name because they were chiefly studied by Muhammedans and Jews.

Bacon was specially interested in light, possibly through the influence of Ibn-al-Haitham. He gave the laws of reflexion, and described mirrors and lenses with the general facts of refraction, including a theory of the rainbow. He invented some mechanical contrivances and described or predicted others, such as self-propelled ships and flying machines. He considered burning glasses, magic mirrors, gunpowder, magnets, artificial gold and the philosopher's stone, in a confused mixture of fact and fiction.

It is probable that Bacon would only be remembered as a magician had not Pope Clement IV told him to write out his work and send it to him. This Bacon did, and from these books we know his achievements. Unfortunately Clement died, and Bacon, deprived of his protection, suffered imprisonment. Much of his work was forgotten; all through the Middle Ages work was done and lost and had to be rediscovered.

THE MIDDLE AGES

In spite of his modern appreciation of mathematics and experiment and his attacks on the Schoolmen, Bacon never cast off the mediaeval habit of mind in philosophy and theology. He was a true man of science, but born too soon.

In modern days Roger Bacon's criticism of Scholasticism would be accounted effective, but it was too much out of tune with mediaeval thought to produce much result at the time. Towards the close of the thirteenth and at the beginning of the fourteenth century, attacks from a philosophic point of view were made by Duns Scotus and William of Occam, who led a revolt against the Scholastic union of philosophy and religion, claiming freedom for both. This dualist view was linked with a revival of nominalism, with its belief in the sole reality of individual things.

In its turn, the new nominalism with its opposition to Scholasticism was banned by the Church, which tried to impose realism as late as 1473. But the work of Scotus and Occam was a severe blow to the dominance of Scholasticism, though in 1879 an Encyclical of Pope Leo XIII re-established the wisdom of Saint Thomas Aquinas as the official Roman Catholic philosophy.

The fourteenth and fifteenth centuries saw the growth of a new mysticism, especially in Germany. Cardinal Nicholas of Cusa, who made advances in mathematics and physics, showed that a growing plant took something from the air, and supported the theory of the Earth's rotation. He maintained that God must be known by mystical intuition and not by reason. Thus Cusa helped in the final overthrow of mediaeval Scholasticism.

We must not think that Scholasticism did only harm in later ages. It is true that it carried Aristotle's philosophy and science forward into a time when they proved an obstacle to the birth of modern science at the Renaissance. But the Schoolmen kept alive in the minds of men the belief in rational order, in cause and effect. As Whitehead says: 'Galileo owes more to Aristotle than appears on the surface...he owes to him his clear head and his analytic mind... the priceless habit of looking for an exact point and of sticking to it when found.'

CHAPTER IV. *THE RENAISSANCE*

The Causes. It has sometimes been said that the Renaissance produced so great a revolution that its causes cannot be fully understood. But it is clear that many accidents and activities coincided at one time to bring it about. The change began in Italy, partly perhaps because the majestic remains of Roman building made it easier for men's minds to recover some of the lost Roman culture, but partly also because in Italy it was the fashion for 'knights and noble ladies' to live chiefly in the towns where intercourse was easy, and not, as in northern lands, on their estates in feudal isolation.

Navigation was facilitated by the magnetic compass, discovered by the Chinese in the eleventh century, and brought to Europe a hundred years later. Early in the fifteenth century the Portuguese began a great period of exploration, discovering the Azores in 1419 and later tracing the west coast of Africa. It became generally accepted that the Earth was a sphere, and that, as the Greek Poseidonius had held, by sailing westward, ships might reach the shores of Asia, and bring back to Europe the rich trade of India and Cathay. With the support of Ferdinand and Isabella of Spain, Columbus sailed from Palos and landed on the Bahamas in 1492. Doubtless these discoveries enlarged not only the known world but also the minds of men. Moreover, besides the direct growth in material resources, when the gold and silver of Mexico and Peru swelled the currency, the consequent rise in prices, as always, stimulated industry and trade and thus increased wealth, giving opportunity for leisure, study and invention.

When a number of factors are at work, the total result at the beginning is only the sum of the separate results, but, when the results overlap, cause and effect act and react, and the whole process becomes cumulative. And so with the material, moral and intellectual factors in the changes of the sixteenth century; somewhat suddenly they combined in the irresistible torrent of the Renaissance.

A literary harbinger appeared in Petrarch (1304–1374), who tried to restore both a taste for good classical Latin in place of the dog-

THE RENAISSANCE

Latin of the Schoolmen, and also the Greek and Roman claim for liberty of the reason.

In the early years of the fifteenth century a growing interest in classical literature drew from the East many Greeks, who, from their modern tongue, were able to teach the ancient language. This influx was quickened by the capture of Constantinople by the Turks in 1453; more teachers came, bringing manuscripts with them, while a search in European libraries disclosed others. Thus Greek again became familiar after a lapse of eight or nine hundred years, and the humanists who first read it played a chief part in the widening of mental horizons which afterwards made science possible. Another important factor in the advancement of learning was the invention, about the middle of the fifteenth century, of printing with movable type, which brought books into many more hands.

Leonardo da Vinci. The influence of personality in past times is often difficult to trace, and it is only since the manuscript note-books of Leonardo da Vinci have been deciphered that we are able fully to appreciate his universal genius. Leonardo (see Plate III, facing p. 29) was born at Vinci between Florence and Pisa in 1452. He lived successively in the courts of Florence, Milan and Rome, and died in 1519 in France, the servant and friend of Francis I.

Leonardo was a painter, sculptor, engineer, architect, physicist, biologist and philosopher, and in all these subjects supreme. Observation and experiment were to him the only true method of science; the opinions of ancient writers could never be conclusive though they might be useful as a starting-point; the most helpful of all were the writings of Archimedes, and for manuscripts of his works Leonardo eagerly sought.

Leonardo grasped the principle of inertia, 'every body has a weight in the direction of its movement'. He understood the impossibility of perpetual motion as a source of power, and deduced therefrom the law of the lever, which he treated as the primary machine. He recovered Archimedes' results in hydrostatics, and dealt also with hydrodynamics—the flow of water through channels and the propagation of waves over its surface, waves in air and the laws of sound, and recognized that light also showed some of the properties of waves.

Some of the fossils found on mountains must have been laid down in sea water; changes must have occurred in the crust of the Earth, and new mountains been thrown up. But no catastrophic happenings are needed, the river Po 'will lay down land in the Adriatic as it has already formed a great part of Lombardy'—a uniformitarian theory three centuries before Hutton.

As a painter and sculptor Leonardo, like Botticelli and Albrecht Dürer, was led to anatomy. In the face of prejudice, he dissected 'more than ten human bodies', making anatomical drawings which are both accurate and works of art. In physiology he explains how the blood carries nutriment to the parts of the body and removes waste products. It almost looks as though he understood the circulation of the blood a hundred years before Harvey. He dismisses scornfully the follies of astrology, alchemy and magic; to him nature is orderly; even in astronomy he holds that the cosmos is a celestial machine, and the Earth a star like the others. All this is markedly different from Aristotle's belief that the heavenly bodies, unlike our Earth, are divine and incorruptible.

Leonardo says that mathematics are concerned with ideal mental concepts and within that realm give certainty. Other sciences should begin with observation, use mathematics if possible, and end with one clear experiment—a good attempt to set forth scientific method. As a philosopher, Leonardo seems to have held an idealistic pantheism, though, with a well-balanced mind, he accepted the essential Christian doctrine as an outward form for his inward spiritual life. But he lived in the brief interval when the Papacy itself was liberal and humanist; later his attitude might have been untenable.

Incidentally he names others, earlier and contemporaneous, who were interested in mathematics and scientific experiment. A circle of kindred spirits evidently lived in Italy, and we know little of them, except from Leonardo's note-books. The Schoolmen prepared men's minds by teaching that the world was understandable, but a new method was necessary; induction from nature had to replace deduction from Aristotle or Thomas Aquinas, and these Italians, above all Leonardo, helped to start the change.

THE RENAISSANCE

The Reformation. Humanism was brought to the north of Europe by students who had worked in Italy. Johann Müller (1436–1476) translated into Latin Ptolemy and other Greek writers, and founded an observatory at Nürnberg. His Ephemerides, the precursors of our Nautical Almanacs, were used by the Portuguese and Spanish explorers. Again, Desiderius Erasmus, chief figure of the northern Renaissance, attacked evils such as monastic illiteracy, Church abuses and Scholastic pedantry, setting himself to show what the Bible really said and meant.

For a few years, culminating with the reign of Pope Leo X (1513–1521), the Vatican was a living centre of ancient culture. But the capture of Rome by the Imperial troops in 1527 broke up this intellectual and artistic life, and soon afterwards the Papacy reversed its liberal policy, and became an obstacle to modern learning. Freedom of thought had to be won through much tribulation by the rough path opened by Martin Luther.

The Reformers had three chief objects: (1) the re-establishment of Church discipline; (2) the reform of doctrine and a return to a supposed primitive simplicity; and (3) a loosening of dogmatic control and a measure of freedom for the individual judgment based on Scripture. The first motive carried the people; the second was an appeal to an older precedent, as in our own day there have been appeals to the 'first four centuries'. But the third object is the one that concerns us here. Although Calvin was as bad a persecutor of free thought as any Roman Inquisitor, he had not the power of the Mediaeval Church behind him, and the disintegration of Christendom, sad though it was from some other points of view, did indirectly help to secure liberty of thought and speech.

Copernicus and Astronomy. Nicolaus Koppernigk (1473–1543), whose father was a Pole and mother a German, was a mathematician and astronomer, and became famous under his Latinized name of Copernicus.

The cosmos as described by Hipparchus and Ptolemy accepted the common-sense view of the Earth as the centre of all things, drawing them to it by gravity, and explained the apparent movement of the heavenly bodies by a series of cycles and epicycles. The system conformed to Aristotle and even to the facts, but it was unpleasantly complicated.

Even while the Scholasticism of Aristotle and Saint Thomas Aquinas was predominant, the form of Neo-Platonism accepted by Saint Augustine survived in Italy as an alternative philosophy. It contained elements derived from the Pythagoreans, the idea that the Earth moved round a central fire, and that the ultimate mystical reality was to be found in numbers; the simpler the relation of the numbers the nearer the truth.

Copernicus spent six years in Italy as a pupil of Novara of Bologna, who criticized the Ptolemaic system as too cumbrous. Copernicus searched what books were available, and found that, according to Cicero, Hicetas thought that the Earth revolved on its axis, and, according to Plutarch, others had held the same opinion or even thought with Aristarchus that the Earth moved round the Sun. Given confidence by this ancient authority, Copernicus found that all the phenomena were simply explained, and so hung together that nothing could be altered without confusing the whole scheme. 'Hence for this reason', he says, 'I have followed this system.'

He takes a sphere of fixed stars beyond all as the frame of the Universe and gravity as a universal force. Of the moving bodies, Saturn completes a circuit in thirty years, Jupiter in twelve, Mars in two, then the Earth, with the Moon going round it in an epicycle, next Venus circling in nine months and finally Mercury in eighty days. 'In the middle of all...the Sun...sitting on a royal throne, governs the circumambient family of stars...we find...a wonderful symmetry in the universe and a definite relation of harmony.' Thus to Copernicus, as to Pythagoras and to Plato, the object was to find the simplest and most harmonious picture of the heavens.

Copernicus printed a short abstract of his work in 1530 and the complete book, *De Revolutionibus Orbium Cœlestium*, was published in 1543 with a Preface by Osiander, suggesting that the theory was only an aid to mathematical simplicity. This led to a mistaken idea that Copernicus himself did not regard it as physical reality. The book also met with criticism based on the science of the time. If the

THE RENAISSANCE

Earth revolved, would not things thrown upward lag behind? Would not loose objects fly away from the ground and the Earth itself disintegrate? Therefore the theory only made its way slowly, and did not become widely known till Galileo, using his newly invented

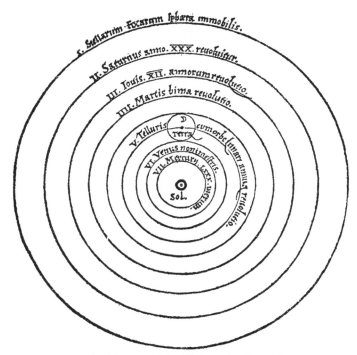

FIG. 8. Diagram of the Universe by Copernicus

telescope, revealed Jupiter's satellites, a solar system in miniature. But by 1616 the Papacy had become alarmed. Galileo was reproved and the theory condemned as 'false and altogether opposed to Holy Scripture', though it might be taught as a mere mathematical hypothesis. But the adverse decision was never ratified by the Pope, and in 1822 the Sun was given formal sanction to become the centre of the planetary system. Galileo's 'persecution' has been exaggerated; he only suffered a mild reprimand and detention and died peacefully in his bed.

Nevertheless the Copernican system brought about a revolution in the minds of men. Instead of the Earth being the centre of the Universe, it became merely one of the planets, and although this change does not necessarily dethrone man from his proud position as the object and summit of creation, it does suggest doubts about the certainty of that belief.

Thomas Digges, an English mathematician and engineer, accepted the Copernican theory, but not Copernicus' fixed stars. He thought the Universe was infinite and the stars scattered through boundless space. This view was also held by Giordano Bruno. But Bruno was a pantheist, and attacked other orthodox beliefs. For this, and probably not for his astronomy, he was condemned by the Inquisition, and burned at the stake in 1600.

The accuracy of astronomical observation, especially of planetary motions, was much improved by Tycho Brahe, a Danish noble of Copenhagen (1546–1601). The year before his death he was joined by John Kepler (1571–1630), to whom he bequeathed his unique collection of data.

Kepler is often represented as solely engaged in searching for laws of planetary motion, and verifying three of them ready for a coming Newton to explain. But Kepler was after nobler game. Saturated with Platonic ideas, he believed that God made the world in accord with the principle of perfect numbers; he was really searching for ultimate causes—the music of the spheres, and the mathematical harmonies in the mind of the Creator.

The three laws he established were: (1) the planets travel in ellipses with the Sun in one focus; (2) the area swept out in any orbit by the straight line joining the centres of the Sun and a planet is proportional to the time; (3) the squares of the periodic times which the different planets take to describe their orbits are proportional to the cubes of their mean distances from the Sun. He was even more pleased by other fancied relations which later observation has not confirmed. About 1590 mathematics were much developed by the

THE RENAISSANCE

invention of symbolic algebra, the chief credit for which belongs to François Viète.

Chemistry and Medicine. It was hoped that the revival of Greek learning would cause the same improvement in medicine as in literature and philosophy. Doubtless, when physicians turned from commentaries to the writings of Hippocrates and Galen themselves, a great increase in knowledge occurred, but when this knowledge had been discovered, accepted and systematized, men came again to rely too much on authority, till once more they began to observe and experiment; medicine then became allied with the chemistry emerging from alchemy to form a school of iatro-chemists or spagyrists.

The Arabs had taken and modified the Pythagorean idea that the primary elements were to be found in principles or qualities and not in substances. They believed that the fundamental principles were those of 'sulphur', that part of a body which made it combustible and vanished on burning, 'mercury' which distilled over as liquid, and 'salt', any solid residue. This theory passed with other Arab learning into Europe.

Theophrast von Hohenheim (1490–1541), a Swiss physician, wandered over Europe studying minerals and mechanical contrivances, and the diseases and remedies of different nations, before practising medicine for a while at Basle, where, from Celsus, the great physician of Roman times, he was given the name of Paracelsus, by which he is better known. He despised most orthodox men of science, and in medicine turned from the authority of Galen and Avicenna to his own observation and experience. As a chemist he prepared many substances, among them ether, and discovered its anaesthetic properties by experiments on chickens, without appreciating its value for mankind.

Van Helmont, mystic and experimenter, born at Brussels in 1577, recognized for the first time that there are different kinds of aeriform substances, and invented the name 'gas' to describe them. Believing that water was the sole element, he planted a willow in a weighed quantity of dried earth and supplied it with water only. At the end of five years the willow had gained in weight by 164 pounds while the earth had lost only 2 ounces. Van Helmont drew the conclusion that the new substance of the willow was made of water only, a legitimate conclusion indeed until, a century later, it was shown that green plants absorb carbon from the carbon dioxide in the air.

Sanctorius (1561–1636) used an improved thermometer to measure the temperature of the human body, and Dubois (1614–1672), better known as Sylvius, applying chemistry to medicine, taught that health depends on the fluids of the body, opposite in kind, combining with each other to form a milder substance—the first theory of chemistry not based on the phenomena of flame, leading afterwards to a general study of acids, alkalies and salts and so to the idea of chemical affinity.

Anatomy, Physiology and Botany. A prejudice against the dissection of human bodies prevented the revival of anatomy till the thirteenth century, and in the fourteenth, after the work of Mondino, it again became stereotyped. The note-books of Leonardo were not made known at the time, and the first modern anatomist of general influence was Andreas Vesalius (1515–1564), who published *Fabrica Humani Corporis* in 1543, setting forth not what Galen or Mondino taught, but what he himself had observed and was prepared to demonstrate. His work on the bones, veins, abdominal organs and brain was specially notable, and he was the first to see the importance of the shape of the skull in the classification of mankind. Before the end of the sixteenth century anatomy was freed from the trammels of ancient authority.

In physiology Vesalius accepted the current ideas that food is endowed in the liver with *natural spirit*, which the heart converts into *vital spirit* and the brain into *animal spirit*; this last 'is a quality rather than an actual thing', and is employed for the operations of the 'chief soul', and for bodily movement 'by means of nerves, as it were by cords.... As regards the structure of the brain, the monkey, dog, horse, cat and all quadrupeds which I have hitherto examined... resemble man in almost every particular.'

Van Helmont held that in plants and brute beasts there is only 'a certain vital power...the forerunner of a soul'. In man, the sensitive soul controls the functions of the body, and acts by means of *archaei* its servants, which work directly on the organs of the body by means of ferments—not chemical ferments, but mystic agencies of another kind. Sylvius also believed in ferments, but his ferments were ordinary chemical agents, like the vitriol which causes effervescence when poured on chalk. This chemical physiology stimulated useful work by Silvius and his pupils.

In considering the functions of the blood, the doctrines of Galen were an obstacle. Galen taught that the arterial and venous blood were two separate tides which ebbed and flowed carrying 'vital' and 'natural' spirits to the tissues. Servetus, physician and theologian, who was burned by Calvin at Geneva for his heterodox opinions, discovered the circulation of the blood through the lungs, but the function of the heart in maintaining the flow through lungs and body was only made clear when William Harvey (1578–1657) was led to 'give his mind to vivisections'.

Harvey was educated at Gonville and Caius College, Cambridge, and after some years abroad, he returned to England and practised as a physician. He was on friendly terms with Charles I, and showed him the development of the chick in the egg. It is said that, during the battle of Edgehill, in charge of the young princes, he sat under a hedge reading a book. With the king he retired to Oxford, and for a time he was Warden of Merton.

In 1628 he published a small but most important volume: *Exercitatio Anatomica de Motu Cordis et Sanguinis*. He points out that in half an hour the heart deals with as much blood as is contained in the whole body, and that therefore the blood must find its way from the arteries to the veins and back to the heart. 'I began to think', he says, 'whether there might not be *a motion as it were in a circle*. Now this I afterwards found to be true'—found that is by observation of the heart as seen in the living animal. Harvey treats the problem as one of mechanics, and solves it as such. Another book, *De Generatione Animalium*, appeared in 1651, and described the greatest advance in embryology since Aristotle.

To complete Harvey's work on circulation, the microscope was necessary. In 1661 Malpighi of Bologna thus discovered in the lung of a frog that the arteries and veins are connected by capillary tubes, and not by structureless flesh as had been supposed. Malpighi also examined microscopically the glands and other organs, and gave the first description of the microscopic changes which appear as the egg starts to produce the living bird.

The Renaissance increased the security of life and led to a simultaneous development in artistic feeling. These factors, with the accompanying increase in wealth, encouraged the laying out of public and private parks and gardens, and the better cultivation of trees, vegetables and flowers. Botanic gardens were founded at Padua in 1545 and afterwards at Pisa, Leyden and elsewhere. Medicine too started gathering grounds and distilleries, where herbs were grown and converted into drugs.

The revival of scientific botany was delayed by the mediaeval doctrine of signatures, whereby the colour or other visible character of a plant was thought to disclose the use for which God designed it. Plant-lore too was mingled with magic and other superstitions.

The first to give accurate accounts of plants with drawings from nature was Valerius Cordus (1515–1544), who thus made the first appreciable advance in systematic botany since Dioscorides fifteen hundred years before. Among botanists of the period mention should be made of Jean and Gaspard Bauhin, who put together extensive catalogues of plants. Other herbals appeared; one was published by William Turner in the years 1551 to 1568, and another, more famous but less accurate, by John Gerard in 1597.

Mediaeval Bestiaries had been compiled, not from nature but from Pliny the Elder. A better spirit was now abroad; animals were studied and zoological gardens founded, one of the earliest at Lisbon.

Magnetism and Electricity. William Gilbert of Colchester (1540–1603), Fellow of Saint John's College, Cambridge, and later President of the College of Physicians, gave in his book *De Magnete* all known facts about magnetism, adding many observations of his

THE RENAISSANCE 57

own. The compass needle, discovered by the Chinese in the eleventh century, was soon used by Muslim sailors, and appeared in Europe a hundred years after its discovery.

Gilbert investigated the forces between magnets, and showed that a magnet, when freely suspended, besides pointing to the north and south, dips in England with its north pole downwards through an angle depending on the latitude. He concluded that the Earth itself

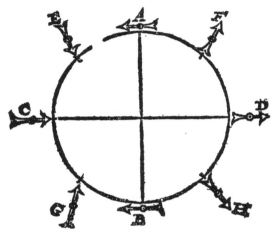

FIG. 9. Diagram of declination of iron magnet

was a magnet, with poles nearly, but not quite, coincident with the geographical poles. For a uniform lodestone, the strength of the magnetism was proportional to its quantity or mass—perhaps the first clear recognition of mass as distinct from weight.

Gilbert also found that when amber was rubbed it showed forces, which he measured with a suspended needle. To describe these results, he coined the name 'electricity' from the Greek word ἤλεκτρον, amber, though he discovered that other substances were also active. He imagined that an aethereal, non-material influence was emitted by the magnet or electrified substance, embracing neighbouring bodies and drawing them towards itself. He used similar half-mystical ideas to explain the motions of the Sun and planets.

Philosophy. Though the great predominance of Aristotelian Scholasticism had been shaken by Duns Scotus, William of Occam and Nicholas of Cusa, it remained very powerful at the time of the Renaissance, and the philosophy and science of Aristotle kept much of their old prestige. When those with the newer outlook wished to experiment in order to test a hypothesis, the orthodox inquirer could meet them with a quotation from Aristotle, and sometimes find his authority still accepted.

It was this position which made so useful the message proclaimed to the world by Francis Bacon (1561–1626), Lord Chancellor of England, 'to extend more widely the limits of the power and greatness of man'. Bacon held that, by recording and tabulating all possible observations and experiments, the relations would emerge almost automatically. But in nature there are so many phenomena and so many possible experiments that advances are seldom made by the pure Baconian method. Insight and imagination must be used at an early stage of the inquiry, and a tentative hypothesis formed by induction from the facts. Then its consequences must be deduced logically and tested for consistency among themselves and for concordance with the primary facts and the results of *ad hoc* experiments. Hypothesis after hypothesis may have to be examined till one passes all the tests, and, for a time at all events, becomes an accepted theory, which can predict phenomena to a high degree of probability. The methods of science are more speculative—more poetical—than Bacon thought. Nevertheless he was the first to consider formally, if inadequately, the philosophy of inductive science, and he supported with statesmanlike eloquence the new experimental method.

René Descartes (1596–1650), who was born in Touraine and also lived in Holland and Sweden, improved the mathematics used in physics and founded modern critical philosophy. For the first time he applied algebra in the methods of co-ordinate geometry. Straight lines OX and OY are drawn from O at right angles. The position of a point P may then be specified by stating the distance OM on x and PM on y. If y increases evenly as x increases, we pass over the diagram in a straight line OP. If y equals x^2 multiplied by a constant, we get

THE RENAISSANCE

a parabola OP' and so on. These equations may be treated algebraically and the results interpreted geometrically to solve many physical problems.

Descartes applied mechanics to construct a theory of the cosmos. He regarded the physical universe as a closely packed plenum with no empty spaces; motion can then only occur in closed circuits, for there is no vacuum into which a body can pass. On these ideas Descartes formed a theory of vortices, in which a stone is drawn to the earth and a satellite to its planet, while the planet and its satellite are whirled in a greater vortex round the Sun. At a later date Newton proved mathematically that the necessary properties of these Cartesian vortices were inconsistent with observation, but they formed nevertheless a bold attempt to reduce the colossal problem of the sky to terrestrial dynamics. In Descartes' scheme God as First Cause started the machine, which then runs spontaneously.

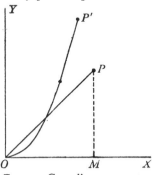

Fig. 10. Co-ordinate geometry by René Descartes

Descartes was the first to formulate complete dualism, with an entire separation of soul and body, mind and matter. Previously the soul was regarded as of the nature of air or fire, and mind and matter differed only in degree and not in kind. Descartes, to whom mind was immaterial and matter was really inert and dead, thus made it possible to consider the human body as a mechanism without banishing the soul altogether, indeed, though they are completely different, thought is as real as matter—*cogito ergo sum*.

Descartes' system was criticized by the nominalist Thomas Hobbes (1588–1679), who would have none of the Cartesian dualism, and held that the only reality was matter in motion. Regardless of difficulties, Hobbes took sensation, thought and consciousness as due to the motion of atoms in the brain, but offered no explanation of how the connexion between two apparently quite different types of activity comes about.

Witchcraft. It is strange that, in the time of the Renaissance, when knowledge of all kinds made a new start in growth and modern experimental science may be said to have begun, there should have been a recrudescence of ancient magic in the form of a belief in sorcery and witchcraft.

When the Christian Church first conquered the world, unconscious of the effect of the mystery religions in the formulation of its belief in earlier times, intelligent men had come to regard magic in its various forms as a relic of paganism which was dying out. So the Church took a lenient view—to call up Satan was not heresy; it was merely sin. But in the later Middle Ages, in Manichaean heresies the Devil became a disinherited Lucifer, an object of worship to the oppressed. Saint Thomas Aquinas set himself to explain away the lax attitude of the early Church, and in 1484 Pope Innocent VIII gave formal sanction to the popular belief in sorcery and witchcraft. Thus was forged a new weapon to combat heretics—they could be declared sorcerers and the fury of the mob roused against them.

After the Reformation the Protestants used the scriptural injunction 'thou shalt not suffer a witch to live' to vie with Romanists in the hunt. On the Continent, where torture was legal, most of those accused confessed. In England, where torture was allowed in Prerogative Courts but not at Common Law, they mostly declared their innocence. Very few men ventured their lives by protesting against the mania. Perhaps the first was Cornelius Agrippa; others were John Weyer, physician to Duke William of Cleves, and Reginald Scot, a Kentish squire who took the modern common-sense view that the whole thing is a mixture of ignorance, illusion and false accusation. A Jesuit, Father Spee, attended nearly two hundred victims to the stake at Würzburg. Horrified at the experience, he declared his belief that they all were innocent, and that anyone could be made to confess by the tortures used. In two hundred years of Europe the number of victims is estimated at three-quarters of a million or more—a disgraceful episode in the history of mankind. But the civilized world discovered that it had ceased to believe in witches before it had stopped burning them. The change was chiefly due to the advance of science, which was slowly defining the limits of man's mastery over nature and the methods whereby it is attained.

CHAPTER V. *GALILEO AND NEWTON: THE FIRST PHYSICAL SYNTHESIS*

Galileo. If we are asked to name the man who did most to start physical science on the triumphant course which lasted for nearly three hundred years, we must answer: Galileo Galilei (1564–1642) (see Plate IV, facing p. 62). Leonardo, Copernicus and Gilbert, each in his different way, foreshadowed the coming revolution, but Galileo went farther, and in his writings we recognize for the first time the authentic modern touch. He brought the theory of Copernicus to the practical test of the telescope, but above all in his work on dynamics he combined observation and induction with mathematical deduction tested by experiment, and thus inaugurated the true method of physical research. In his work there is no reliance on an authoritative and rational scheme as in Scholasticism; each problem is faced in isolation, each fact is accepted as it stands with no desire to make it fit into a universal pre-ordained whole; concordance, if possible at all, comes slowly and partially; mediaeval Scholasticism was rational, modern science is in essence empirical, accepting brute facts whether they seem reasonable or not.

In 1609 Galileo heard a rumour that a Dutchman had made a glass that magnified distant objects. From his knowledge of refraction, he was able at once to do likewise, and soon made a telescope which magnified thirty diameters (see Plate V, facing p. 63). Surprising discoveries followed: the surface of the moon, instead of being perfect and unblemished as held by Aristotle, was seen to be broken by rugged mountains and valleys: countless stars, flashing into sight, solved the problem of the Milky Way. Round Jupiter four satellites revolved, a model of the Earth and its moon moving round the Sun as taught by Copernicus on *a priori* grounds of mathematical simplicity. But the Professor of Philosophy at Padua refused to look through the telescope, and his colleague at Pisa laboured with logical arguments before the Grand Duke to refute Galileo.

Galileo's most original and important work was the establishment of the science of dynamics on a combined experimental and mathematical basis. Stevinus of Bruges had considered the inclined plane and the composition of forces, and thus made some advance in statics, but men's ideas on motion were a confused medley of Aristotelian theory and disconnected observation. It was thought that bodies were essentially light or heavy, and therefore rose or fell to find their 'natural places'. Stevinus, in an experiment afterwards confirmed by Galileo, let fall a heavy body and a light body together, and showed that they struck the ground practically simultaneously. In repeating the experiments more accurately, Baliani, a Captain of archers at Genoa, rightly referred the slight difference to the friction of the air. Since the forces on the two bodies are measured by their weights, there must be some other quantity which resists the setting in motion, and the experiments show that in each body it must be proportional to the weight. This quantity, as we saw above, was recognized also by Gilbert; it is what we call mass, and later it was explicitly discussed by Newton.

Copernicus and Kepler proved that the motion of the planets could be described in mathematical terms. Galileo thought that bodies on the Earth 'in local motion' might also move mathematically. Watching the bronze lamps, hanging from the roof of the cathedral at Pisa, he saw that the swings, whether large or small, occupied equal times, and so discovered the principle of the pendulum. A falling body moves with increasing speed; what is the law of the increase? Galileo first tried the hypothesis that the speed was proportional to the distance fallen through; but this, he found, involved an inconsistency, and he then tried another hypothesis, that the speed varied with the time of fall. Galileo deduced its consequences to compare with experiment. The speeds being too great for his instruments to measure, he used inclined planes, having found that a body falling down such a plane acquired the same velocity as though it had fallen through the same vertical height. His measurements agreed with his second hypothesis and its mathematical consequence that the space described increases with the square of the time.

Galileo's inclined planes gave another important result. After running down one plane a ball will run up another to the same vertical height, and, if the second plane be horizontal, the ball will

Plate IV GALILEO

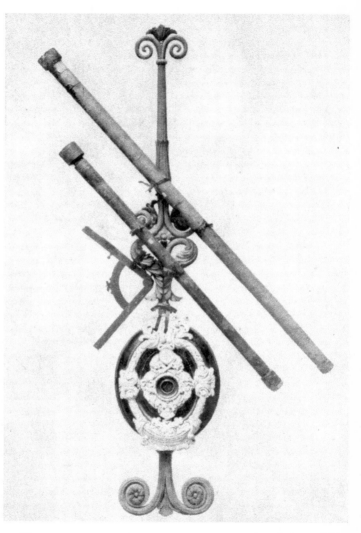

Plate V   GALILEO's original telescope

THE FIRST PHYSICAL SYNTHESIS

run along it in a straight path till stopped by friction. Thus a moving body continues to move in virtue of inertia till some opposing force comes into play. It had been supposed that a force was necessary to maintain motion; the planets had to be kept going by Aristotle's Unmoved Mover, or Kepler's action of the Sun exerted through an aether. Galileo's result showed that, when the planets have been set in motion, they only need a force to pull them away from a straight path, and keep them swinging round the Sun in their orbits. The way was opened for Newton.

Galileo also began the study of the strength of materials, the bending of beams and other problems in elasticity, and carried out further work on hydrostatics. Also, with the newly invented glass tube, he made a thermoscope, in which the expansion of air in a bulb measured the temperature. The production and manipulation of glass tubes did much for experimental science.

Galileo, like Kepler, looked for mathematical relations, not however in a search for causes, due to mystic numbers, but in order to discover the laws by which nature works, caring nothing 'whether her reasons be or be not understandable by man'—a contrast to homocentric Scholasticism, in which the whole cosmos is made for man. To Aristotle space and time were somewhat unimportant and vague 'categories'. Galileo, after his experiments on falling bodies, gave them that primary and fundamental character which Newton adopted and passed on to his successors.

Galileo distinguished between primary qualities, such as extension or shape, which, he held, cannot be separated from a body, and secondary qualities, such as colour or smell, which, following Democritus, he referred to the senses of the percipient, the observed body being merely 'atoms and a void'. Galileo accepted the atomic theory, discussing speculatively how differences in the number, weight, shape and velocity of atoms may cause differences in taste, smell or sound. Galileo's time and space, with his atomic theory and its consequences, led later to dualism and materialism in other men, while some have referred to Galileo's results many of our present philosophic discontents. But it may equally be said that Galileo revealed and clarified the difficulties obscured or concealed by Aristotle. Galileo confessed that he knew nothing about the cause of gravity or the origin of the Universe, and declared it better 'to pronounce that wise, ingenuous and modest sentence, "I know it not"'.

Boyle, Huygens and others. The philosophy of Hobbes, founded on the science of Galileo, did not escape criticism. The Cambridge Platonists pointed out that a theory which made extension and its modes the only real properties of bodies could not explain life and thought. They tried to reconcile mechanical views with religion by an apotheosis of space. Malebranche identified infinite space with God himself. Spinoza held that God is the immanent cause of a pantheistic Universe, and the Cartesian dualism of mind and matter is resolved in a higher unity when viewed *sub specie aeternitatis*. Thus contemporary philosophers tried to escape from their growing difficulties by invoking the power of God.

Prevalent theories were well discussed by Sir Kenelm Digby (1603–1665), who ridiculed Aristotle's 'essential qualities', and held with Galileo that all things could be explained by particles 'working by means of local motion'. Again, Galileo's new mathematical methods were clearly set forth in the lectures of Isaac Barrow, Newton's teacher at Cambridge. Space and time are absolute, infinite and eternal, because God is omnipresent and everlasting. This seems to be the first definite formulation of the ideas of absolute time and space as held by Newton.

Stevinus had re-opened the subject of hydrostatics by calculating the pressure of a liquid at a given depth; Robert Boyle (1627–1691), chemist, physicist and philosopher, proved that air is a material substance having weight, its volume being inversely proportional to the pressure, a relation known as Boyle's Law, rediscovered by Mariotte. Boyle observed the effect of a change in atmospheric pressure on the boiling point of water, collected many new facts in electricity and magnetism, and explained heat as a 'brisk' agitation of particles. As a chemist he distinguished element, mixture and compound, prepared phosphorus, collected hydrogen in a vessel over water, though he called it 'air generated *de novo*', and studied the form of crystals as a guide to chemical structure. He dealt with the

THE FIRST PHYSICAL SYNTHESIS

chemistry of common things without reference to the still surviving half-mystical theories, indeed he rejected both the old idea of four elements, and also the 'principles' or 'essences' of sulphur, mercury and salt. In 1661 he published *The Sceptical Chymist...touching the experiments whereby Vulgar Spagirists*[1] *are wont to endeavour to evince their Salt, Sulphur and Mercury to be the True Principles of Things*. In a trialogue, Boyle's spokesman explains that he still feels doubts about elements or principles, 'notwithstanding the subtile reasonings I have met with in the books of the Peripatetiks,[2] and the pretty experiments that have been shew'd me in the Laboratories of Chymists'. Gold can be alloyed or dissolved in *aqua regia* and yet recovered in its original form; this suggests unalterable atoms of gold surviving various combinations rather than vague Aristotelian elements or Spagyrist principles. Thus Boyle had grasped the ideas on which modern chemistry was afterwards founded.

Boyle accepted the view that secondary qualities are mere sensations, but he justly pointed out that 'there are *de facto* in the world certain sensible and rational beings that we call men', so that their sensations, and with them the secondary qualities of all bodies, are part of the world in being and therefore as real as the primary qualities. The mechanical world and the thinking world (we may say) are both parts of the whole problem which philosophy has to face. God made the world in the beginning and (Boyle holds) His 'general concourse' is continually needed to keep it in being and at work. This partial return to the Indian and Arabic idea of continual creation and re-creation is also the physical aspect of the Christian doctrine of immanence.

Thus Boyle was an interesting philosopher as well as a distinguished physicist and chemist. His diversity is described in an Irish epitaph, which, it is said, called him 'Father of Chemistry and Uncle of the Earl of Cork'.

Blaise Pascal (1623–1662), best known as a theologian, was also an experimental physicist. He arranged for a barometer, newly

---
[1] See p. 53.
[2] Aristotle's School was called Peripatetic from the custom of master and pupils walking together in the garden of the Lyceum at Athens.

invented by Torricelli, to be carried up the Puy de Dôme, when the height of the mercury column became less, clearly because the pressure of the atmosphere fell, and not through nature's 'abhorrence of a vacuum' as taught by the Aristotelians. Pascal was also the originator of the theory of probability, which, beginning with games of chance, has proved to be of great and still increasing importance in philosophy, science and social statistics. Indeed the basis of all empirical knowledge may be said to be a matter of probability expressible in terms of a bet.

The most striking advances in dynamics since the work of Galileo were made by Christiaan Huygens (1629–1695), who published his book *Horologium Oscillatorium* in 1673. Assuming the constancy in moving bodies of *vis viva*, which we now call kinetic energy, he deduced the theory of the centre of oscillation and opened a new method in mechanics. The work done by a force f moving a body through a space s is fs, and Huygens found that this is equal to the *vis viva* produced, that is $\frac{1}{2}mv^2$. He proved also the relation between the length of a pendulum and its time of swing, and dealt with circular motion. If a body of mass m describes with velocity v a circular path of radius r, by Galileo's result a force must act towards the centre. Huygens showed that, in modern language, the acceleration is v^2/r, so that the force is mv^2/r. Newton must have reached the same conclusion in 1666, but did not publish it till a later date. Huygens also improved the telescope and discovered Saturn's rings.

Scientific Academies. In tracing the intellectual environment of Newton and his contemporaries, it is necessary to take into account the societies or academies which were then being formed. The new learning, striving to push its way against the opposition of the Aristotelians, only slowly entered the Universities of Europe. But the numbers interested in 'natural philosophy' were growing rapidly, and here and there they met together for discussion. At Naples in 1560 was formed an *Accademia Secretorum Naturae*; from 1603 to 1630 an *Accademia dei Lincei* existed in Rome, and in 1651 the *Accademia del Cimento* was founded by the Medici at Florence.

In England a society began to meet in 1645 at Gresham College

THE FIRST PHYSICAL SYNTHESIS

or elsewhere in London, under the name of the Philosophical or Invisible College. In 1648, owing to the Civil War, most of its members moved to Oxford, but in 1660 it returned to London, and in 1662 it was incorporated by a Royal Charter of Charles II as 'The Royal Society of London for Promoting Natural Knowledge'. In France a corresponding *Académie des Sciences* was founded by Louis XIV in 1666, and similar institutions followed in other lands.

The influence of these bodies in facilitating discussion and making known the results of research has had much to do with the rapid growth of science, especially as most of them soon began to issue periodic publications. An independent *Journal des Savants* appeared at Paris in 1665, and three months later it was followed by the *Philosophical Transactions* of the Royal Society, at first a private venture of its secretary.

Even in the second half of the seventeenth century, survivals of mediaeval thought, especially Aristotelian Scholasticism, contended with the new science based on mathematics and experiment. In mechanics, after the work of Galileo, the newer outlook was learning to express its results in terms of matter and motion. This was helped by a revival of the atomic theory, accepted by Galileo, and developed by Gassendi. The tendency was hastened by Huygens, though echoes of old controversies were still heard. Boyle in 1661 thought it worth while to argue against both Scholastic and Spagyrist concepts in chemistry, and both alchemy and astrology were still taken seriously. When Newton went to Cambridge and was asked what he meant to study, he is said to have answered: 'Mathematics, because I wish to test judicial astrology.' At a later date he possessed a chemical laboratory, and it is probable that he spent much more time there labouring fruitlessly at alchemy, as well as at more hopeful chemistry, than he gave to the dynamics which changed the whole outlook of astronomy.

The Greek idea of an aether had been used by Kepler to explain how the Sun kept the planets moving, by Descartes as a fluid in which to form his vortices, by Gilbert in his theories of electricity and magnetism, and by Harvey to enable the Sun to send heat to the heart and blood of living animals. But interplanetary aether was still confused with Galen's aethereal or psychic spirits, used by the mystic to explain the nature of being. The modern distinction between mind and matter, soul and body, had not become clear. The 'soul', the 'animal spirits' and similar concepts were regarded as 'vapours' or 'emanations', things to us material. A unity was thus maintained by most men except Descartes, who was the first to see an essential difference between inert matter and the thinking mind. The usual line seems to have been drawn between solids and liquids on one side, and air, fire, aether and spirit on the other. Also nearly all men of science and philosophers in the middle of the seventeenth century looked on the world from the Christian standpoint; the idea of antagonism between religion and science is of a later date. Even Hobbes, who was a philosophic materialist and defined religion as 'accepted superstition' agreed that it should be established and enforced by the State. Hobbes was exceptional; almost all men adopted the fundamental theistic belief, to which, they thought, any theory of the cosmos to be true must conform.

Newton and Gravitation. Isaac Newton (1642–1727), *qui genus humanum ingenio superavit*, is still accepted as the bearer of the most illustrious name in the long roll of science [see frontispiece]. Born at Woolsthorpe in Lincolnshire on Christmas Day 1642, the year in which Galileo died, he entered Trinity College, Cambridge, in 1661, and attended the mathematical lectures of Isaac Barrow. He was elected as a Scholar of the College in 1664 and as a Fellow in 1667. In 1665 and the following year, driven to Woolsthorpe by plague at Cambridge, he meditated on planetary problems.

For in those days [he says] I was in the prime of my age for invention, and minded Mathematics and Philosophy more than at any time since.

It is one of the ironies of history that, after those years, Newton did his best to avoid being pushed into work on 'Mathematics and Philosophy' or into publication of his results.

Galileo's work had shown the need of a force acting towards the Sun to keep the planets circling in their orbits. Newton is said to have seen the clue while idly watching the fall of an apple in the

Woolsthorpe orchard. He wondered about the cause of the fall, and the distance the apparent attraction of the Earth for the apple would reach: would it perchance reach the Moon, and explain that body's continual fall towards the Earth away from a straight path?

From Kepler's third law Newton deduced that the forces keeping the planets in their orbits must be, at all events approximately, inversely as the squares of their distances from the centre about which they revolve, and thereby compared the force needed to keep the Moon in her orbit with the force of gravity on the surface of the Earth, and, says Newton, 'found them answer pretty nearly'. But, always averse to publication, he took no step to make his discovery known: perhaps to 'answer pretty nearly' was not good enough.

But the most probable cause of Newton's delay was that pointed out in 1887 by J. C. Adams and J. W. L. Glaisher. The sizes of the Sun and planets are so small compared with the distances between them that the whole of each may fairly be treated as concentrated in one place. But compared with the size of the apple or its distance from the ground, the Earth is gigantic. The problem of calculating the combined attraction of all its parts was one of great difficulty. Newton solved it in 1685 when he proved that a uniform sphere of gravitating matter attracts bodies outside it as though all its mass were concentrated at the centre. This justified the simplification whereby the Sun, planets, Earth and Moon were taken as massive points. Glaisher says:

> No sooner had Newton proved this superb theorem—and we know from his own words that he had no expectation of so beautiful a result till it emerged from his mathematical investigation—than all the mechanism of the universe at once lay spread before him...it was now in his power to apply mathematical analysis with absolute precision to the actual problems of astronomy.

The way was then clear for Newton to deal with his old problem of the apple and the Moon. Using a new French measurement of the Earth's size due to Picard, he found concordance within narrow limits, and the proof of identity was complete: the familiar fall of an apple to the ground and the majestic sweep of the Moon in her orbit are due to one and the same unknown cause.

Meanwhile the question of gravity was under general discussion, especially at the Royal Society. If the planetary orbits, really ellipses, are taken as circles, it follows from Huygens' results and Kepler's third law that the force must be inversely proportional to the square of the distance.[1] But the real paths are ellipses, and no one in London seemed able to solve the actual problem. Halley therefore went to see Newton at Cambridge, and found that he had worked it out two years before, but had lost his notes. However he drafted another solution and sent it to Halley with 'much other matter'. Urged by Halley, who paid for the publication, Newton wrote out his work, and in 1687 issued the *Principia*, the 'Mathematical Principles of Natural Philosophy', the greatest book in the history of science.

The heavenly bodies, to Aristotle divine, incorruptible, and different in kind from our imperfect world, were brought by Newton within range of man's inquiry, and shown to move in accordance with the dynamical principles established by terrestrial experiments. Their behaviour could be deduced from the one assumption that every particle of matter attracts every other particle with a force proportional to the product of the two masses and inversely proportional to the square of the distance between them. The movements so calculated were found to agree accurately with the observations of more than two centuries. Halley proved that even comets move in accord with gravity. He calculated the times of appearance of the comet named after him, indicating that it was the same as the comet pictured in the Bayeux tapestry which was thought to presage disaster to the Saxons in 1066.

Newton also investigated the attraction of the Sun and Moon on the waters of the Earth, and placed tidal theory on a sound footing, though the complications introduced by land and channels still defy complete explanation.

The proofs in the *Principia* are given in geometrical form, but it is possible that Newton first obtained some of them by new mathematical methods which he had himself devised. The chief of these was his method of fluxions—finding the rate of change of one variable

---

[1] Kepler's third law states that the squares of the periodic times, and therefore r^2/v^2, are proportional to r^3. Hence v^2 varies as $1/r$, and v^2/r, which by Huygens' proof gives the acceleration and therefore the force, is proportional to $1/r^2$.

x with another y. Newton wrote this rate as $\dot{x}$, but Leibniz, who invented an equivalent method, apparently independently, used the better notation dx/dy, which was adopted in the further development of the differential calculus. But Newton seems to have had an amazing power of seeing the solution of a problem by intuition; sometimes his proofs may have been merely helps to weaker minds.

Mass and Weight. The concept of mass as inertia, distinct from weight, is implicit in the work of Galileo. It appears explicitly in the writings of Gilbert and of Baliani, who refers both to *moles* and *pondus*. Newton defined mass as 'the quantity of matter in a body as measured by the product of its density and bulk', and force as 'any action on a body which changes, or tends to change, its state of rest, or of uniform motion in a straight line'. He then summarizes his results in three laws of motion, the second stating that change of motion, which, in modern terms, means rate of change of momentum, is proportional to the moving force.[1]

No serious criticism of this formulation appeared till Mach in 1883 pointed out that Newton's definitions of mass and density involve a logical circle; he defines mass in terms of density, while density can only be defined as mass per unit volume. But it is possible to avoid this difficulty. If two bodies act on each other, as by mutual gravitation or by a coiled spring joining them, the ratio of their opposite accelerations depends only on something in the bodies which we may if we please call mass. We can then define the relative masses of the two bodies as the inverse ratio of their accelerations, and the force between them as the product of either mass and its own acceleration. The logical circle is thus avoided.

We can also escape from it in another way. We have ideas derived from experience about space or length and about time; our muscular sense similarly gives us the idea of force. Equal forces, as roughly measured by this sense, are found to produce unequal accelerations on different pieces of matter, and the inertia of each piece, its resistance to the force f, may be called its mass, and may be defined as the inverse of the acceleration produced by a given force, or

[1] The rate of change of velocity is called acceleration, and the rate of change of momentum, mv, is therefore $m\alpha$, so that force is mass multiplied by acceleration.

$m = f/\alpha$. This may bring psychology into physics, but it is interesting to note that it is possible to avoid a logical circle in physics by doing so. Next we find by experiment that the relative masses of two bodies as thus defined are roughly constant; then we can make the hypothesis that this rough constancy is rigorously true, or better, true to a high degree of accuracy, and use M as a third fundamental unit to those of length L and time T. All the innumerable deductions from this hypothesis of constancy proved exactly true till recent days.

Mass being settled, there remains the problem of its relation to weight, the force by which a body is drawn towards the Earth. The answer is implicitly given by the experiments on falling bodies carried out by Stevin, Galileo and Baliani, which showed that a heavy and a light body fall together, that is, their accelerations are the same and equal to about 16 feet per second in every second of time. If W_1 and W_2 be the weights, we have by experiment $\alpha_1 = \alpha_2$, that is $\dfrac{W_1}{m_1} = \dfrac{W_2}{m_2}$ or $\dfrac{W_1}{W_2} = \dfrac{m_1}{m_2}$, the weights are proportional to the masses. Newton verified this result more accurately by showing that pendulums of the same length have the same time of swing, whatever be their mass or material. Since weight hastens the swing and mass retards it, Newton's result shows that weight and mass must be proportional. He pointed out how surprising it is that gravity thus proceeds from a cause depending not on the surface but on the 'quantity of the solid matter' the bodies contain; gravity must penetrate to the centres of the Sun and planets.

Mach also pointed out that the dynamical work of Galileo, Huygens and Newton only means the discovery of one fundamental result. In his experiments on falling bodies, Galileo found that the velocity increased with the time, $v = \alpha t$, so that $mv = ft$, that is, the increase in momentum is measured by the product of the force and the time, which gives the Newtonian law $f = \dfrac{mv}{t} = m\alpha$. Now half the final velocity is the average velocity, and this multiplied by the time gives the space traversed, so that Galileo's $v = \alpha t$ or Newton's $mv = ft$ becomes Huygens' $\tfrac{1}{2}mv^2 = fs$, which states that the *vis viva*, our kinetic energy, is equal to the work done (p. 66). As Mach said, only one fundamental principle had been discovered.

THE FIRST PHYSICAL SYNTHESIS

But if Galileo had happened first on the fact that the square of the velocity increased with the space traversed, or $v^2 = 2\alpha s$, he would have thought that $\frac{1}{2}mv^2 = fs$, Huygens' equation of work and energy, was the primary and important relation. It was merely the accident of history which gave that place to force and momentum, and caused delay in the use of the concepts of work and energy.

Optics and Light. The law of refraction, that the sines of the angles of incidence and refraction bear a constant ratio, had been discovered by Snell in 1621, while the colours seen in the rainbow and in cut glass were, of course, familiar to men. In 1666 Newton 'procured a triangular glass prism to try the celebrated phenomena of colours', and his first paper, published in 1672, was on light.

Newton showed that white light is made up of light of various colours, differently refracted by passing through the prism, the most refracted being violet and the least red. On these results he solved the problem of the rainbow, and explained the colours which disturbed the vision through the then known refracting telescopes, but he concluded erroneously that colour could not be prevented without at the same time destroying the refraction which gave magnification. To evade the difficulty he invented a reflecting telescope.

He examined the colours of thin plates, well known in bubbles and other films. By pressing a glass prism on to a lens of known curvature, the colours were formed into circles which he could measure, since called 'Newton's rings'. Using light of one colour only, the rings became alternately light and dark; it was then clear that the colours with white light were due to the abstraction of one colour after another in turn. He also repeated and extended experiments made by Grimaldi, who had found that narrow beams of light are bent at the edges of obstacles, making the shadows larger than expected and showing fringes of colour. Newton proved that the bending is increased by passing the light through a narrow slit, and made careful observations and measurements. He also carried further Huygens' experiments on the double refraction of Iceland spar, and pointed out that they proved that a ray of light cannot be symmetrical but must be different on its different sides.

In considering the nature of light, another discovery, made by

Roemer, was important. When the Earth is between the Sun and Jupiter, the eclipses of his satellites appear about fifteen minutes earlier than when the Earth is beyond the Sun, and the light has to cross the Earth's orbit. Thus light does not travel instantaneously, but needs a finite time.

The nature of light has always been a subject of wonder and speculation. The idea that light consists of particles can be traced back to the Greeks—particles either coming from the visible object or projected from the eye to feel the object, though Aristotle held that light was action in a medium. Descartes thought it was a pressure transmitted through his plenum, and Hooke suggested that it was a rapid vibration in a medium, a theory worked out in some detail by Huygens.

Newton, in dealing with the passage of heat through a vacuum, asked in his queries:

Is not the Heat convey'd through the Vacuum by the Vibrations of a much subtler Medium than Air... and exceedingly more elastick and active?

But the fact that normally light travels in straight lines made Newton think that primarily it must consist of projected particles, though to explain the other properties he had to suppose that the particles stirred up vibrations in the medium which conveys radiant heat—vibrations which put the particles into 'Fits of easy Reflexion and easy Transmission' alternately. Thus he gave to his particles some of the properties of waves.

Now it is a remarkable fact that, after a century of a pure wave theory of light and forty years of electric particles, physicists are now finding that these particles are associated with waves of extremely short wave length (see p. 152). Indeed this modern view of electron particles with their trains of waves is very much like Newton's inspired guess. His amazing insight into nature is once more demonstrated.

Newton fitted up a laboratory in the garden behind his rooms in Trinity College, and there spent much time in chemical and alchemical experiments. He wrote no book or even paper on this subject; all that remain are his manuscript notes, and some of the queries at the end of his *Optics*. He seems to have been chiefly interested in metals, in chemical affinity and in the structure of matter. Thus he states that the most fusible alloy of lead, tin and bismuth contains those metals in the proportion of 5 : 7 : 12. In query 31 he writes:

> When Salt of Tartar runs *per deliquium* is not this done by an Attraction between the Particles of the Water which float in the Air in the form of Vapours? And why does not common Salt, or Saltpetre, or Vitriol, run *per deliquium*, but for want of such an Attraction?

Newton and Philosophy. Newton carried farther Galileo's method of scientific investigation. Postponing the question of *why* things happen, he concentrates on the problem of *how*. This is clearly seen in his work on gravity. For although Newton was often said to have established 'action at a distance', in fact he regarded such an idea as absurd. He proved that bodies moved *as though* particles attracted each other, but he clearly and often states that he does not know why, or by what mechanism, gravity works. That is a separate problem which comes at a later stage of the inquiry. Indeed it may never be solved, for science deals with appearance and not necessarily with reality.

Again the same distinction is seen in Newton's queries about chemical affinity:

> Have not the small Particles of Bodies certain Powers, Virtues or Forces, by which they act... upon one another for producing a great part of the Phaenomena of Nature?... How these Attractions may be perform'd I do not here consider. What I call Attraction may be perform'd by impulse, or by some other means unknown to me.

The first step in scientific research, as carried on by Galileo and Newton, is to examine known facts and frame a hypothesis to reduce them to order—the process of induction. The logical consequences of the hypothesis must then be deduced by mathematics or otherwise and compared with observation or experiment. If they agree, the hypothesis may be called a theory, and used to suggest further inquiry, observation and experiment, for, as has been well said, the larger becomes the sphere of knowledge, the greater grows its area of contact with the unknown.

Newton's dynamics and astronomy involve the ideas of absolute space and time, though we can see now, in our era of relativity, that those ideas do not necessarily follow from the phenomena. Newton also, like Galileo, accepted the atomic theory, though the time had not yet come to put it in the definite numerical form framed by Dalton a century later.

The work of Galileo and Newton, which went so far to express dynamical and even some physical phenomena in terms of time, space and moving particles of matter, undoubtedly suggested, when translated into philosophy, a mechanical or materialist creed, and, as we shall see later, this was its outcome in the eighteenth century, especially in France, where it helped forward first deism and then atheism. But to Newton and his contemporaries that conclusion would have been quite foreign. Firstly, Newton himself, as shown above, clearly distinguished between a successful mathematical description of natural phenomena and a philosophical explanation of their causes. Secondly, Newton and his immediate followers regarded the new discoveries as a revelation of the power and wisdom of God. Newton wrote:

> This most beautiful System of the Sun, Planets and Comets could only proceed from the counsel and dominion of an intelligent and powerful Being...God...endures for ever and is everywhere present, and, by existing always and everywhere, he constitutes duration and space.

Thus to Newton God is not only a First Cause, but is also immanent in Nature. Newton sums up:

> All these things being considered, it seems probable to me that God in the Beginning formed matter in solid, massy, hard, impenetrable, moveable particles, of such Sizes and Figures, and with such other Properties, and in such proportion to space, as most conduced to the End for which he form'd them.

During his early years in Cambridge, while he 'minded Mathematics and Philosophy', Newton studied mystical writings and much theological literature as well as treatises on alchemy, which often

THE FIRST PHYSICAL SYNTHESIS

showed mystical leanings. His theological opinions were unorthodox, and, though he was a Fellow of his College, he never took Holy Orders. But it is clear that he had a firm belief in God, and felt the utmost confidence that his scientific work went to confirm that belief.

Newton in London. Newton played an important part in defending the University against the attack on its independence by James II. He was elected to the Convention Parliament which settled the succession of the Crown, and was again elected in 1701.

In 1693 he suffered from a nervous breakdown, and his friends decided that it would be well for him to leave Cambridge. They obtained for him the post of Warden of the Mint, and later he became Master, the highest office there. He gave up his chemical and alchemical researches, and put the papers concerning them into a locked box.

His move to London marked a complete change in his life. His scientific achievements won for him a pre-eminent position, and for twenty-four years, from 1703 till his death, he was President of the Royal Society, which gained much authority by his unique powers and reputation. In spite of the absence of mind which marked his early years, his work at the Mint showed that he had become an able and efficient man of affairs, though he was always nervously intolerant of criticism or opposition.

His niece Catherine Barton, a witty and beautiful woman, kept house for him, and it was on this second part of his life that the eighteenth century built up its Newtonian legend. Catherine married John Conduitt; their only child became the wife of Viscount Lymington and the Lymingtons' son succeeded to the Earldom of Portsmouth. Thus Newton's belongings passed into the possession of the Wallop family. See below.[1]

[1] In 1872 the fifth Lord Portsmouth gave some of Newton's scientific papers to the Cambridge University Library. At a later date some of his books and papers were sold. Part of the papers were acquired by Lord Keynes; the books were bought by the Pilgrim Trust, and have now (1943) been presented to Trinity College.

CHAPTER VI. *THE EIGHTEENTH CENTURY*

Philosophy. Some writers of the time of Newton and after, writers who were primarily philosophers, either touched explicitly on science, or, at all events, dealt with branches of philosophy with scientific implications. Of these men John Locke (1632–1704), though most of his life fell in the seventeenth century, belonged in spirit to a later age, and may well serve to introduce it.

Locke practised as a physician; he wrote against Scholastic ideas in medicine and in favour of observation as used by his friend Dr Sydenham. But Locke's chief work was his *Essay Concerning Human Understanding*, in which he argues that thoughts are due to experience, either sensations of external things, or reflexions on the operations of our minds. We know nothing of substances except their attributes, and those only from sense impressions. When these attributes show themselves in a constant connexion, we gain the idea of an underlying substance, the existence of which in some form seems to Locke a fair inference from our success in co-ordinating appearances. To fix abstract ideas by means of words involves danger, because the meaning of words may change—an early criticism of language. Locke also began introspective psychology, watching steadily the operations of his own mind. He concluded that knowledge is the discernment of agreement, either of our thoughts among themselves or with external phenomena; but these external relations can only be established by induction from specific instances, and a knowledge of nature can only be an affair of probability, liable to be upset by new facts. In political, philosophic and religious thought, Locke upheld a moderate and rational liberalism. He insisted on the toleration of various religious opinions—in that age a great proof of originality.

Newton's science and Locke's philosophy together led George Berkeley, Bishop of Cloyne, to accept the Newtonian picture of the world, but to point out that it is only a picture of the world revealed by the senses, and only the senses which make it real; the material world is made real only by being apprehended by some living mind.

God must exist, because the material world to be a real world in the absence of human minds needs to be continually realized and regulated by Him.

To the plain man this seems a denial of the existence of matter. It has led to much criticism, good and bad, beginning with Dr Johnson, who thought he could refute Berkeley by kicking a stone, and continuing till it inspired a recent writer of Limericks. But it does seem true that *the world we know* is only made real by the senses; we cannot know the world of reality which may or may not lie beneath, though, with Locke, we may think its existence a fair inference from the knowledge of appearance revealed by the senses.

David Hume (1711–1776) went farther and denied reality to mind as well as to matter; all that he left real was a succession of 'impressions and ideas'. He argued that the empiricists, in appealing exclusively to sense experience, made it impossible to pass to the inductive inference of universal laws. Hume held that to look on one happening as the cause of another is merely an instinct, perhaps founded on coincidence and unwarranted. Both Locke and Hume regarded metaphysical reality as beyond the reach of human reason. Hume also separated reason and faith, as the late mediaevalists revolted against the rational synthesis of Scholasticism.

Scepticism as regards the possibility of a knowledge of reality was held also by Kant (1724–1804) and extended to all the principles of science and philosophy. The world of science is the world revealed by the senses, the world of appearance, not necessarily the world of ultimate reality. Leibniz thought that pure reason could unfold external, unchanging truth; Kant set himself to save as much of Leibniz's pure reason as Hume had left standing. Kant maintained we could still believe that the moral sense is as real as the starry heaven, indeed more real, the one form in which reality discloses itself to the human mind.

Leaving the writings of the professional philosophers, we turn to the influence of the Newtonian system on the general thought of the eighteenth century, especially in France, where it found its way into the famous *Encyclopédie*. As we have seen, Newton and his friends regarded his picture of the Universe as a revelation of the glory of God, and biologists such as Ray echoed their words; but it must be confessed that Newton's work produced a very different effect on Voltaire and those of a like way of thinking. Their general wave of sceptical religious thought got a powerful reinforcement from Newton's success in explaining the mechanism of the heavens, and from the possibility of extending similar mechanical principles to other branches of science. Naturally they exaggerated: Laplace conceived a mind able to foretell the progress of nature for all eternity if but the masses and their velocities were given. Few would be so bold nowadays, when a principle of indeterminacy is hovering over the perplexed thoughts of physicists. But Voltaire wrote

It would be very singular that all nature, all the planets, should obey eternal laws, and that there should be a little animal, five feet high, who, in contempt of these laws, could act as he pleased, solely according to his caprice.

Voltaire ignored the problems of the meaning of natural laws, of the functions of life and of the human mind, and the possible scope still left for free will. But doubtless he was expressing vividly the current French opinions about the philosophic and religious bearings of Newtonian science. Laplace told Napoleon that he had no need of the hypothesis of a Creator, though Lagrange made the comment that it was a good hypothesis, it explained many things. In England, where the eighteenth-century Church was tolerant, and men more used to holding simultaneously beliefs at the moment apparently incompatible, the mechanical outlook never became so common as in more logical lands. Newton's countrymen for the most part accepted both Newton's science and their accustomed religion; the idea of antagonism only became prevalent in the nineteenth century.

Meanwhile in France a current of popular thought was running towards materialism, a philosophy easily understood, as old as the Greek Atomists and recently expounded by Hobbes. The word materialism is often used in a loose sense to mean atheism, or indeed any view we dislike. But here it is used in its strict meaning—a belief that dead matter in the form of solid impenetrable Newtonian particles (or perhaps of their modern derivatives) is the sole ultimate

THE EIGHTEENTH CENTURY 81

reality; that thought and consciousness are but by-products of matter.

The allied, but not identical, theory of mechanical determinism was also taken over by the French materialists, while Holbach argued that, since man, a material being, thinks, matter itself is capable of thought. This seems an assumption of the very thing to be explained, a mere restatement of the problem. For rough, everyday work, and for investigating the details of science, an assumption of materialism is useful, but there is danger that it should be taken as the philosophy of science as a whole, and gain the prestige which the success of science gives. In fact, matter, like all other concepts of science, is only known to us through its effect on our senses, and we are brought back to the problem of knowledge. Even in the eighteenth century, Locke, Berkeley and Hume had shown that materialism, at all events in its then form, should have failed to satisfy. We shall meet materialism again, especially in Germany, during the nineteenth century.

Mathematics and Astronomy. A controversy about priority between Newton and Leibniz caused an unfortunate separation between English and Continental mathematicians. The Englishmen, using either geometry or Newton's method of fluxions, did comparatively little towards developing the differential calculus and its consequences, which were carried forward abroad by James Bernouilli, Euler and others. But useful results were obtained in England by Taylor and Maclaurin in the expansion of certain mathematical series.

Newton's work was made known in France especially by Maupertuis, while Voltaire wrote a popular account. For a time France was the chief centre of scientific activity. Lagrange (1736–1813) created the calculus of variations, and systematized the subject of differential equations—work which proved of use in physical and astronomical problems. In his treatise *Méchanique Analytique* he treated the whole of mechanics on the principles of conservation of energy and virtual velocity—what is gained in power is lost in speed. Maupertuis gave the name of 'action' to the sum of the products of space (or length) and velocity, and showed that light always went by the path of least action. We shall meet an allied form of 'action' in recent quantum physics.

The Newtonian system was extended by Pierre Simon Laplace (1749–1827), who, beginning as the son of a cottager, skilfully guided himself to end as a *Marquis* of the Restoration. He treated problems of attraction by Lagrange's method, and proved that the solar system was stable, perturbations of one planet due to other planets or comets being only temporary and correcting themselves. He framed a nebular theory, whereby the system of Sun and planets was formed from a rotating mass of incandescent gas, an idea which had also occurred to Kant. Modern investigation shows that the theory fails for the solar system, but may be true for the much larger aggregates of stars seen in formation in spiral nebulae. Newton had shown that the velocity of a wave should be equal to the square root of the elasticity concerned divided by the density. Taking the usual elasticity of air, this formula gives too small a figure for the velocity of sound. Laplace traced the discrepancy to heat, which, developed in the wave by the sudden compression and absorbed by the following expansion, increases the elasticity of the air and therefore the velocity of the sound.

Gravitational astronomy has followed the lines laid down by Newton and Laplace. An apparently final test was given in 1846 by the successful prediction by J. C. Adams and Leverrier of the presence of an unknown planet, afterwards called Neptune, from the perturbations of another, Uranus. The success of Newton's formulation is astonishing; it is only the most powerful modern apparatus that can show certain minute discrepancies in favour of Einstein. But, to a very high degree of accuracy, Newton's work stands; it is absorbed in something wider but not superseded. It was completed in another direction in 1781 by Henry Cavendish (1731–1810), who used Michell's torsion balance to measure the attraction between two balls, and thus determine the gravitational constant.

Kepler's observations gave a model of the solar system, but the scale of the model was not known till some one distance had been measured in terrestrial units. We shall give recent values in a later

THE EIGHTEENTH CENTURY

chapter; but fairly good results were found by Richer in 1672 (p. 88). Among other results was an estimation of the velocity of light by Bradley in 1729 from the aberration of the stars as the Earth moves in its orbit.

Chemistry. In the early years of the eighteenth century Homberg, linking with the ideas of Sylvius, studied the reactions of acids and alkalies to form salts—work which afterwards led to theories of chemical structure. In 1732 Boerhave of Leyden published the best chemical treatise of the time and thus helped to co-ordinate existing knowledge.

Most early work on chemistry sprang from a desire to explain the phenomena of flame. When bodies are burnt, it seems that something escapes. This something, for long called sulphur, was given the name of 'phlogiston', the principle of fire, by Stahl (1660–1734), physician to the king of Prussia. Rey and Boyle had shown that when metals were burnt the solid increased in weight; thus, as Venel pointed out, phlogiston must possess a negative weight—a return to Aristotle's idea of a body essentially light. Chemistry, ignoring the achievements of physics, learned to express itself in terms of phlogiston, which dominated the chemical thought of the latter part of the century.

Meanwhile many new substances were discovered. Oxygen had been obtained from saltpetre by Borch in 1678, and it was again prepared and collected over water in 1729 by Hales, who still thought it was air modified by the presence of some other substance. But about 1755 Joseph Black of Edinburgh discovered that a new ponderable gas, distinct from air, was combined in the alkalies. He described it as 'fixed air'; it was what we now call carbon dioxide or carbonic acid. In 1774 Scheele discovered chlorine. Joseph Priestley (1733–1804) prepared oxygen by heating mercuric oxide, and rediscovered its unique power of supporting combustion and respiration. Cavendish demonstrated the compound nature of water in 1781, thus finally banishing it from the list of elements, though he still called its constituent gases phlogiston and dephlogisticated air.

Thus we see men of the eighteenth century collecting chemical facts and slowly feeling their way forward under the hampering restriction of false assumptions and theories, till there came the hour and the man. Antoine Laurent Lavoisier (1743–1794), who was guillotined with other farmers of the taxes because 'the Republic had no need of savants', had repeated the experiments of Priestley and Cavendish, weighing accurately his reagents and products. He found that, though matter may alter its form and properties in a series of chemical actions, its total amount as measured by the balance is unchanged. The constituents of water, which Lavoisier named hydrogen and oxygen, were gases with the ordinary properties of mass and weight. The concept of phlogiston with its negative weight vanished from science, and the principles of Galileo and Newton were carried over into chemistry.

Physiology, Zoology and Botany. The physiology of breathing and its connexion with combustion had been studied in the seventeenth century, specially by Boyle, Hooke, Lower and Mayow. They proved that air is not homogeneous, but contains an active principle —*spiritus nitro-aereus*—needed both for breathing and burning, our modern oxygen. Metals when burned increase in weight by the absorption of this spirit. Lower showed that the change in the colour of blood from purple to red took place not in the heart but in the lungs, where the blood came into contact with air and absorbed its vital particles. Much of this sound work was forgotten till the chemical facts and their interpretation were rediscovered in the eighteenth century by Priestley and Lavoisier.

We have seen that Sylvius threw over van Helmont's spiritualistic ideas of the control of the body by a sensitive soul working through *archaei*, and tried to explain bodily functions by effervescences of the same kind as that produced when vitriol is poured on to chalk.

But the pendulum now swung the other way. Stahl maintained that changes in the living body were fundamentally different from ordinary chemical reactions, being governed by a sensitive soul on a plane above physics and chemistry. Sensation and its concomitants are modes of motion directed by the sensitive soul. Stahl was a forerunner of modern vitalists, though they found it necessary to convert his very definite 'sensitive soul' into a much vaguer 'vital principle'.

These physiological questions were discussed in *Institutiones Medicae* (1708) by Boerhave, one of the greatest physicians of modern times. Blood pressure was first measured by Hales, who also after Ray investigated the pressure of the sap in trees.

One new practice introduced in the eighteenth century is important, among other reasons, because it foreshadowed modern methods of securing immunity from specific diseases. In 1718 Lady Mary Wortley Montagu introduced from Constantinople the practice of inoculation for small-pox with pus from a light case, and thus began the control of one of the most prevalent and deadly of diseases. Towards the end of the century, Benjamin Jesty acted on the common belief that dairymaids who suffered from the lighter cow-pox did not catch small-pox. Apparently independently, Edward Jenner, a country doctor at Berkeley, investigated the subject scientifically, and devised the method of vaccination, whereby the benefits of immunity are obtained without the risks involved in the older process of inoculation from small-pox itself.

It has been said that the year 1757 marks the dividing line between modern physiology and all that went before, because in that year was published the first volume of the *Elementa Physiologiae* of Albrecht von Haller (1708–1777), the eighth and last volume appearing in 1765. In this great work Haller gives a systematic account of the then state of physiological knowledge. He himself studied the mechanics of respiration, the development of the embryo and muscular irritability. From observation of disease and experiments on animals, he concluded that the nerves alone feel; they are therefore the only instruments of sensation, as, by their action on the muscles, they are the only instruments of motion. All nerves are gathered into the *medulla cerebri* in the central parts of the brain, so it too must feel and present to the mind the impressions which the nerves have collected and brought. Von Haller thought the nerves were tubes containing a special fluid, and that, since sensation and movement have their source in the medulla of the brain, the medulla is the seat of the soul.

About this time there was an increase in the number of animals obtained for royal menageries, and Buffon (1707-1788) wrote a comprehensive Natural History of Animals. He regarded a classification which placed man among animals as 'une vérité humiliante pour l'homme', but said that had it not been for the express statements in the Bible, one might be tempted to seek a common origin for the horse and the ass, the man and the monkey.

The microscope threw new light on the structure and functions of the organs of animals, and also revealed the existence of vast numbers of minute living beings both animal and vegetable. In ancient and mediaeval times, men believed that living things might arise spontaneously, frogs, for instance, by the action of sunshine on mud. But doubts had been raised by Redi, who showed that, if the flesh of a dead animal were protected from insects, no grubs or maggots appeared in it. Redi's work was repeated by the Abbé Spallanzani (1729-1799), who moreover proved that not even minute forms of life appeared in decoctions first boiled and then protected from the air.

The application in the seventeenth century of the microscope to the study of plants, especially by Grew and Malpighi, began to give correct ideas about their structure. In 1676 Grew recognized the stamens as the male organs, referring the original idea to Sir Thomas Millington. Others followed, making it clear that, in the absence of pollen from the stamens, no fertilization of the ovum or setting of seed is possible.

A recent biography by Dr C. E. Raven has made clear the important part played by John Ray (1627-1705), botanist, zoologist, the first scientific entomologist and a writer on the relations between religion and science. Ray gave up his Fellowship and other offices at Trinity College, Cambridge, because, though he had not signed the Covenant and remained a Churchman, he did not approve the Act of Uniformity. He left Cambridge and retired to his birthplace, Black Notley in Essex. He travelled through the counties of England and through parts of Europe, often with Francis Willughby, studying and making lists of the plants and animals. He also wrote many books, among them a *History of Plants*, a *History of Insects* and *Synopses*

THE EIGHTEENTH CENTURY

of animals, birds, reptiles and fishes. In his books on plants he paid special attention to their medicinal qualities.

In flowering plants the first leaves growing from the embryo, since termed cotyledons, are either single, as in grasses and lilies, or double, as in trees and many other plants. Ray recognized the importance of this difference between monocotyledons and dicotyledons as dividing flowering plants into two great groups, and much improved classification in other ways, using all characters—fruit, flower, leaf, etc.—in order to group together plants with real natural affinities. As regards animals, Ray not only studied their external characters but also their comparative anatomy. He examined fossils, and concluded that they 'were originally the shells or bones of living fishes and other animals bred in the sea'—an enlightened view at a time when fossils were generally regarded either as due to a 'plastic force' or as deposited by Noah's deluge.

He rejected magic and witchcraft, and all superstitious explanations of occurrences, holding to natural causes as revealed by observation. Ray's philosophy is shown in his book *The Wisdom of God in the Works of Creation*, in which, as a biologist, he supports the view of Newton and his disciples as astronomers. Together they succeeded in banishing the idea, continually recurring from Augustine to Luther, that nature is on a plane irrelevant if not hostile to religion, its beauty a temptation, its study a waste of time. But Ray writes:

> There is no occupation more worthy and delightful than to contemplate the beauteous works of nature and honour the infinite wisdom and goodness of God.

In systematic botany Ray was followed by Linnaeus (Carl von Linné, 1707–1778), who framed a convenient scheme of double nomenclature and founded a system of classification on the sex-organs of plants, a system that endured till replaced by the modern plan which, returning in the light of evolution to the ideas of Ray, tries to group plants in their natural relations by a study of all their characters. Linnaeus also dealt with the varieties of men, placing them with apes, lemurs and bats in an order of 'Primates', and dividing them into four groups by skin colour and other differences.

It was partly Ray's work which inspired Gilbert White (1720–1793), naturalist, who, after a long study of animals and plants round his home in Hampshire, published in 1789 a famous literary English classic—*The Natural History of Selborne*. These two, Ray and White, helped the development of our English love of the country and its life which shines so clearly from Gilbert White of Selborne to Edward Grey of Fallodon.

Unfortunately at this time there was little intercourse between men of science and practical gardeners and breeders, who by hybridization and selection were producing new varieties of plants and animals. To horticulturalists the spontaneous appearance of large variations was well known. In animals Bakewell improved by selection the old long-horned cattle and Leicester sheep, while the brothers Colling established the famous shorthorn breed of cattle. The knowledge of these men would have been of great scientific value to biologists.

Geography and Geology. In the seventeenth and eighteenth centuries systematic exploration of the world began to be put on a scientific basis. Apparently the first man sent on a voyage to carry out definite astronomical measurements was Jean Richer, who went to Cayenne in French Guiana in 1672–3 under the auspices of Colbert, the famous Minister of Louis XIV, 'to make astronomical observations of utility to navigation'. The chief problems set were the movements of the Sun and planets, refraction and parallax. From the known relative distances, these observations would give the absolute distances, and the most striking of Richer's results was the revelation of the enormous sizes of the Sun and the larger planets and the dimensions of the solar system itself: the Earth and man on it shrank into insignificance. These measurements, together with Picard's new and more correct estimation of the Earth's diameter, did much to increase the accuracy of astronomical knowledge.

Some years later, William Dampier (1651–1715), after early voyages, went to Jamaica and, for a time, perforce sailed with the buccaneers. He crossed the Pacific Ocean and came home by way of Sumatra and Madras. He kept a Journal, and in 1697 published a Book of Voyages which had a phenomenal success. In 1698 he

THE EIGHTEENTH CENTURY

was commissioned by the Admiralty to command the ship *Roebuck* in exploring 'New Holland', that is Australia. The ship was small and in bad condition; it sank off the island of Ascension on the way home. Nevertheless this was the first voyage planned in England for the purpose of scientific exploration. Dampier not only recorded accurate observations on physical geography, flora and fauna, but he also increased existing knowledge of hydrography, meteorology and terrestrial magnetism.

Dampier's books of voyages set the fashion for a flood of travel literature, beginning with Defoe's *Robinson Crusoe* and Swift's *Gulliver's Travels*. The voyages of Dampier, Cabot and others helped intellectual development, especially in France, and led to writings about island utopias, *Le Bon Sauvage* and other such fancies. In England the story of the Garden of Eden also contributed to the growth of the idea that men of old were better than now, and the 'noble savage' a finer being than civilized man—ideas false indeed, but leading later to the scientific study of anthropology.

Thus interest in exploration steadily increased, and, at the instance of the Royal Society, James Cook (1728–1779) was sent out in 1768 by the Admiralty in command of the ship *Endeavour* to observe a transit of Venus at Tahiti in the South Pacific, having on board two men of science, Banks[1] and Solander. Cook's later voyages of discovery, undertaken in the hope of finding an antarctic continent, failed in this object, but resulted in other information of scientific value, as well as in new knowledge of the coasts of Australia and New Zealand and of the Pacific Ocean. In the earlier voyage, thirty men died out of eighty-five, chiefly from scurvy; in the latter only one died of disease out of 118, a tribute to the success of Cook's researches on anti-scorbutic remedies.

Meanwhile the science and art of navigation were much improved. The determination of latitude is fairly simple by observation of the mid-day altitude of the Sun, but longitude could only be found when the position of the Moon could be predicted by Newton's lunar

---

[1] Afterwards Sir Joseph Banks, a man of fortune and an enthusiastic systematic botanist, who filled the Presidential Chair of the Royal Society for forty-two years, a longer term than that of anyone else in its history.

theory, and only made easy and accurate when John Harrison improved the chronometer in 1761–2 by compensating the effect of changes in temperature by the unequal expansion of two metals. Greenwich time could then be taken on board each ship and compared with astronomical events.

All this growing knowledge of the Earth led naturally to a study of its structure, and to speculation about its history, in fact, to Geology. The large collection of John Woodward, left in 1728 by him to Cambridge University, did much to show, as already held by Leonardo, Ray and others, that fossils were animal and vegetable remains. In 1674 Perrault had proved that the rain falling on the ground was enough to explain the flow of springs and rivers, and Guettard (1715–1786) pointed out how much observed weathering changed the face of the Earth. But the facts were still forced into conformity with Biblical Cosmogonies involving cataclysmic origins by fire or flood.

James Hutton, a Scottish landowner and farmer, after studying land at home and abroad, published in 1785 a *Theory of the Earth* in which he showed that the stratification of rocks and the embedding of fossils were still going on in sea, river and lake. 'No powers are to be employed that are not natural to the globe, no actions to be admitted except those of which we know the principle', a precept known as Hutton's Uniformitarian Theory.

After Hutton, William Smith assigned relative ages to rocks by noting their fossilized contents, and Cuvier reconstructed extinct mammals from fossils and bones, comparing existing animals with fossil forms, thus showing that the past as well as the present must be brought into account in studying the development of living creatures.

Machinery. The most important practical invention of the eighteenth century was the steam-engine, which in the nineteenth revolutionized transport. The steam-engine was produced and improved by empirical methods; it cannot be claimed that its invention was due to the application of scientific principles. The first practical stationary steam-engine was due to Newcomen; it was improved by

Smeaton and principally used for pumping in the Cornish mines. The chief drawback to these early models was the waste involved in cooling the steam-cylinder at each stroke, and the steam-engine only became reasonably efficient when James Watt in 1777 used a separate condensing cylinder kept always cold, and soon after converted linear into rotary motion by a crank shaft. Several primitive paddle-boats foreshadowed the steam-ships of the next century. Land transport was made much easier by better roads, on which fast stage coaches and well-made waggons were replacing the saddle and pack-horses of earlier times.

Building was facilitated by measurement of the strength of materials; processes for bleaching and dyeing improved in the light of growing chemical knowledge and weaving machinery was much developed. The great changes in agriculture were chiefly due to the enclosures going on, whereby mediaeval methods of joint cultivation passed into modern systems of rotation of crops carried out on separate farms. Ploughs were improved, a primitive form of reaper tried, and a drill for sowing corn in rows designed by Jethro Tull.

This summary, of course, is not meant to be complete; it deals only with a few of the more important advances. During the eighteenth century the arts of life were developed in countless directions, but the changes, though great, were not revolutionary; the face of England was little altered.

CHAPTER VII. *PHYSICS AND CHEMISTRY OF THE NINETEENTH CENTURY*

The Scientific Age. The greatest difference between the nineteenth century and those that came before, at least in the subjects of the present book, is to be sought in the change in the relative position of science and industry. Hitherto invention and other improvements in the arts of life had proceeded for the most part independently of science, or set the pace for science to follow and problems for science to solve. The instances where science led the way and practice followed, as, for example, in improvements in navigation, are few in number; but in the nineteenth century they become numerous. A scientific knowledge of electricity led to the electric telegraph, Faraday's experiments in electro-magnetism to the dynamo and the great industry of electrical engineering, and Maxwell's electromagnetic equations, after fifty years' experiment, to wireless telephony and broadcasting. These instances are but examples taken from one department of science; they could be multiplied almost indefinitely; the nineteenth century was the beginning of the scientific age.

Science itself made great advances, especially in physics and chemistry. In their various branches the explanations of new discoveries fitted together, giving confidence in the whole, and it came to be believed that the main lines of scientific theory had been laid down once for all, and that it only remained to carry measurements to the higher degree of accuracy represented by another decimal place, and to frame some reasonably credible theory of the structure of the luminiferous aether. But from 1895 onwards X-rays, electrons, quanta and relativity produced a complete revolution in physical science. It became clear that Newtonian and nineteenth-century physics were only valid within certain limits of size and velocity; below atomic dimensions and at velocities approaching that of light, the former ideas failed, and a wider generalization became necessary, in which the older scheme appeared as a special case, applicable to large-scale phenomena and slowly moving masses.

THE NINETEENTH CENTURY

To construct a quantitative science, definite units are necessary. The multiplicity of weights and measures, by which Britain is still afflicted, was replaced in France by a logical decimal system. Beginning in 1791, the measurements and legislation were finished by 1799, and the system made compulsory as from 1820, not perhaps with complete success.

The unit of length, the metre, was meant to be the ten millionth part of an earth quadrant, but practically it is the distance between two marks made on a certain metallic bar. The kilogram, meant to be the weight of a cubic decimetre of water, is now a mass equal to that of a platinum-iridium standard made in 1799. The second is defined as the 1/86,400 part of the mean solar day. Although the first two are now known to be not exactly what was meant, they are near enough for most practical purposes.

About 1870 it was agreed to use for scientific measurements the centimetre (one hundredth part of a metre), the gram (one thousandth part of a kilogram) and the second; this is usually known as the C.G.S. system of units.

From these units others are derived. That of velocity is one centimetre per second, with dimensions length divided by time L/T, acceleration L/T^2, force ML/T^2, the unit being that needed to give the mass of one gram unit acceleration and called the dyne, while energy is force multiplied by length, ML^2/T^2, the C.G.S. unit being called the erg. Electric and magnetic units also were built up by Gauss on this basis.

Heat and Energy. The scientific concept of heat is derived from our sense-perception, and the thermometer enables us to define a scale with which to measure its intensity. Galileo invented the first thermometer and Amontons first used mercury. Different scales were introduced by Fahrenheit, Réaumur and Celsius. The idea of heat as a quantity was suggested by observation in distilleries, but it was Joseph Black (1728-1799) who cleared up the still existing confusion between heat and temperature, calling them quantity and intensity of heat. He studied the change of state from ice to water and water to steam, finding that, in each change, much heat was absorbed with no rise in temperature; heat, as he said, was rendered latent. He explained the different amounts of heat needed by different substances to produce the same rise in temperature by assigning to each substance a 'specific heat', and showed how to measure quantities of heat in a calorimeter—a vessel containing a known weight of water with a thermometer immersed in it.

It is evident that these experiments needed a theory of heat laying stress on the idea of a quantity which remained constant as the heat passed from one body to another. Although Cavendish, Boyle and Newton had regarded heat as a vibratory motion of the particles of substances, that view did not lend itself, before the days of the theory of energy, to the idea of a quantity remaining constant. It was better to take the alternative hypothesis that heat was a subtle, invisible, weightless fluid, passing freely between the particles of bodies, and this caloric theory served well till the middle years of the century. Sometimes science can be advanced better by a false hypothesis which meets the immediate needs of the time.

The calorists explained the heat developed by friction by supposing that the filings or abrasions, or the main body after friction, had a less specific heat, so that heat was, as it were, squeezed out. Though the American Benjamin Thompson, who in Bavaria blossomed forth as Count Rumford, had shown in 1798 by experiments on the boring of cannon that the heat developed was proportional to the work done, and had no relation to the amount of shavings, his work was forgotten or ignored, and the caloric theory flourished.

But by 1840 it had become apparent that some of the powers of nature were mutually convertible. Assuming that when air is compressed all the work appears as heat, J. R. Mayer calculated the mechanical equivalent of a unit of heat. Sir W. R. Grove, Judge and man of science, wrote on the Correlation of Physical Forces, and von Helmholtz (1821–1894), the great German mathematician, physicist and physiologist, published *Ueber die Erhaltung der Kraft*. The idea of correlation was in the air.

From 1840 to 1850 James Prescott Joule (1818–1889) was engaged in measuring experimentally the heat liberated by mechanical and electrical work. He found that, however the work was done, the expenditure of the same amount of work produced the same quantity of heat. To warm one pound of water through one degree Fahrenheit needed about 772 foot pounds of work—a figure afterwards corrected to 778.

A troublesome double meaning in the word 'force', which had been pointed out by Young, was cleared up when Rankine and William Thomson (Lord Kelvin, 1824–1907) used the word 'energy' in a specialized sense to denote the power of doing work, and measured, if the transformation is complete, by the work done. Joule's experiments showed that the total amount of energy in an isolated system is constant, the quantity lost in work reappearing as heat. Thus the somewhat vague 'correlation of forces' became the quite definite 'conservation of energy'.

It is well to compare this result with the constancy of mass in Newtonian dynamics and in the chemistry of Lavoisier. When we strive to bring order into a welter of unco-ordinated phenomena, to start, that is, a new science, quantities that have this character of permanence inevitably obtrude themselves in our minds. We use them to frame our explanatory scheme, as mass is used in dynamics and chemistry. In studying the relations of heat and work within the limits of the older physics, another quantity emerges, chiefly because of its constancy. We call it energy; use it in our investigations, and laboriously and doubtfully Joule rediscovers its constancy. And so we learn to recognize the conservation of energy as the first law of thermodynamics and place it alongside the conservation of mass. This was a great discovery, but it had its dangers; some men believed that matter was indestructible and eternal; the amount of energy in the Universe constant and immutable. The principles were converted from safe guides for a few steps in empirical science into metaphysical dogmas of doubtful validity. We shall see in the sequel reasons to suppose that a new outlook on these subjects has become necessary.

If heat is a form of energy, accepted from Joule's work, it must consist in motion of the molecules. This kinetic theory, given in early forms by Bernouilli, Waterston and Joule, was first adequately expounded by Clausius in 1857.

The chances of molecular collision will produce molecules moving in all directions with all velocities, the average value of which was calculated by Joule; for hydrogen it is 1844 metres (more than a mile) a second, and for oxygen 461 metres. But Maxwell and Boltzmann applied to the velocities a law of error derived by Laplace and by Gauss from Pascal's theory of probability. The number of molecules

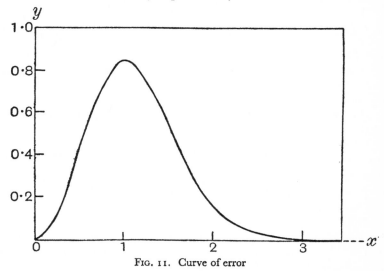

FIG. 11. Curve of error

moving within a certain range of velocity is shown by the diagram. The horizontal ordinate x denotes the velocity, and the vertical ordinate y the number of molecules moving with it. It will be seen how fast the numbers fall off as we pass from the most probable value where the numbers are greatest. This curve of error and its modifications are now of great use in many branches of science. It is impossible to predict the velocity of a particular molecule or the length of life of an individual man, but, with a sufficient number of molecules or men, we can deal with them statistically, and say how many will move within certain velocities, or how many will die in a given year—statistical determinism but individual uncertainty.

Maxwell and Boltzmann also showed that, on the kinetic theory,

THE NINETEENTH CENTURY

the total energy of a molecule should be divided equally among its degrees of freedom, that is the number of co-ordinates needed to specify its position and condition. From this it follows that the ratio of the specific heats of a gas at constant pressure and at constant volume is 1·67 for three degrees of freedom and 1·4 for five—figures confirmed experimentally for gases with monatomic and diatomic molecules respectively.

Boyle's law, which states that the volume of a gas varies inversely as the pressure, may be written as
$$pv = RT,$$
if temperature is also brought in. At high pressures or low temperatures gases depart from this ideal relation. Molecular attraction, which depends on the square of the density, or the inverse square of the volume, will convert p into $p + \frac{a}{v^2}$, and the volume b occupied by the molecules themselves will reduce the varying volume to $v - b$. Thus Van der Waals in 1873 obtained the equation
$$\left(p + \frac{a}{v^2}\right)(v - b) = RT.$$

About 1869 Andrews investigated the continuity of the gaseous and liquid states, and showed that above a certain critical temperature, characteristic of each gas, no pressure, however great, will cause liquefaction. Liquefaction then is a problem of cooling the gas below its critical point. With this knowledge, gas after gas was condensed, till Dewar liquefied hydrogen in 1898, and Kamerlingh Onnes liquefied helium, the last gas to surrender, in 1908.

Thermodynamics. The study of heat and energy was naturally extended to elucidate the laws of heat-engines. In 1824 Sadi Carnot, the son of the 'Organiser of Victory' in Republican France, imagined the simplest ideal case—a frictionless engine in which there is no loss of heat by conduction, etc. The engine must be supposed taken round a complete cycle, so that the working substance, steam, air, or whatever it be, is brought back to its initial state. Carnot's theory was put into modern form by Clausius and William Thomson. The latter showed that all such ideal engines must have the same efficiency, independently of the form of the engine or the nature of the working substance, and depending only on the temperatures T of the source and t of the condenser. We may therefore use an imaginary ideal engine to define temperature, and say that the ratio of T to t is measured by the ratio of H to h, where H is the heat absorbed from the source, and h that given up to the condenser in the ideal engine.

In this way Thomson obtained a thermodynamic or absolute scale of temperature. If all the heat is converted into work, t becomes zero, and we get an absolute zero of temperature, below which nothing can be cooled. He and Joule together, improving on an experiment carried out by Gay-Lussac in 1807, showed that when a gas expands without doing work the change of temperature is very small. This proved that the ideal scale of temperature was practically the same as the scale of the air or hydrogen thermometer, which can easily be constructed.

When changes are going on between heat and other forms of energy, even in ideal conditions heat passes from hot to cold bodies, and, in actual conditions, there are irreversible losses by conduction and radiation as well. These observations can be put in various forms to express a second law of thermodynamics: e.g. 'heat cannot of itself pass from a colder to a warmer body'. It follows that, in an isolated system, heat becomes less and less available for the performance of useful work. This process was named the dissipation of energy. It was put in a different but equivalent form by Clausius, who showed that a mathematical function he named entropy was always tending to increase, and when it reached a maximum an isolated system must be in equilibrium. From these results it was argued that finally all the energy of the Universe would run down into heat uniformly distributed, and no further conversion into other energies be possible in a dead Universe. This is a very big extension of the thesis, and we shall see later that recent discoveries have somewhat modified it. Moreover, even at the time one theoretical way of escape was found by James Clerk Maxwell (1831–1879), who pointed out that energy could be reconcentrated by a being, whom he called a demon, with faculties fine enough to follow and

THE NINETEENTH CENTURY

control individual molecules. By opening or shutting a frictionless trap-door, he (or she) could collect fast moving molecules on one side and slow ones on the other, and thus re-create a difference of temperature.

Thermodynamics were carried farther by Willard Gibbs (1877) with equations which showed that, in a system not isolated but isothermal, that is kept at constant temperature, equilibrium is reached when another function, which he called the thermodynamic potential, is a minimum.

By the use of these two functions, entropy and thermodynamic potential, together with an equation connecting latent heat with temperature, pressure and volume, it is possible to put the whole theory of physical and chemical equilibrium on a sound basis, the equilibrium between different phases being described by a Phase Rule.

The Wave Theory of Light. Thomas Young (1773-1829) overcame some of the difficulties in the way of regarding light as waves in an aether. He studied the coloured or light and dark bands produced by the interference of two overlapping parts of a beam of light passed through two pin-holes in a screen. If one part has to travel half a wave-length farther than the other, the crest of one wave will coincide with the trough of the next, and the wave be there destroyed. From the dimensions of the apparatus and of the interference bands, he calculated the wave-lengths, and found them to be about the one fifty thousandth part of an inch, very small compared with the sizes of ordinary obstacles. A mathematical investigation shows that for such small waves it is only elements in the direct path that produce an effect; the light travels in straight lines except at the edges of shadows.

Hooke had suggested that the waves of light might be transverse to the direction of the beam, and this idea was taken up by Fresnel (1788-1827). The vibrations of ordinary light in a plane at right angles to the rays are complex, but in plane polarized light they are linear, and if a polariscope lets through one such ray it will not let through one polarized at right angles. Fresnel, Green, MacCullagh,

Cauchy, Stokes and others developed the wave theory to a high degree, getting remarkable concordance between theory and experiment.

In order to carry the waves, a medium or aether had to be invented, and, to explain transverse vibrations, that aether must have rigidity. Many elastic solid theories of the aether followed, the chief difficulty being to accept at the same time rigidity and the absence of resistance to the motion of the planets. But when Maxwell showed that light was an electro-magnetic wave, the aether ceased to be necessarily mechanical.

Spectrum Analysis. Galileo and Newton showed that our mechanical laws hold good throughout the bodies of the solar system. To complete the proof of identity, it was necessary to show similarity in structure and composition—an apparently hopeless task.

In 1752 Melvill found that light from flames tinged with metals or salts when passed through a prism gave spectra of characteristic bright lines, and in 1823 Herschel suggested that such lines could be used as a test for the presence of the metals. This led to a long series of observations, in which spectral lines were mapped and recorded. Including the discovery of new elements, spectrum analysis has yielded many important results, culminating with the evidence on which rest our modern theories of atomic structure.

In 1802 Wollaston discovered that the luminous band of the Sun's spectrum was crossed by a number of dark lines. In 1814 Fraunhofer rediscovered and mapped them carefully. Foucault, and independently Bunsen and Kirchhoff, showed that, if complete white light be passed through a flame containing sodium, the dark line called D appears—explaining its presence in the solar spectrum. The intense light from within the Sun, passing through the cooler envelope, loses the sodium light. Many elements known to us on the Earth have thus been shown to occur in the Sun and stars. Conversely helium was first discovered in the Sun and afterwards found in a rare mineral on the Earth. As Stokes pointed out, the dark lines are an example of resonance; a system will absorb vibrations in tune with those it will itself emit. If a source of sound or light is approaching the observer, the frequency is increased, if receding it is lowered. This, called the 'Doppler' effect, is used to measure stellar velocities in the line of sight. The visible spectrum is only a small part of solar radiation. There are long infra-red heat-waves and short ultra-violet ones with intense chemical activity.

Spectra can be formed by gratings as well as by prisms. A grating is made by ruling a number of parallel scratches on a glass plate or on a metal mirror. At the right angle, the scratches cut out all but appropriate waves, thus producing a spectrum.

Maxwell showed that light should exert a minute pressure on a surface on which it falls, and this pressure has been demonstrated experimentally; it is of great importance in cosmical physics. It also allows thermodynamics to be applied to radiation, showing that a black body should radiate in proportion to the fourth power of the absolute temperature, as previously found practically by Stefan. From these results, theoretical and practical, the temperature of the surface of the Sun can be estimated by measuring the rate of emission of heat.

Electric Currents. During the eighteenth century, many experiments were made on statical electricity obtained by friction, and its identity with lightning was demonstrated by Benjamin Franklin and others. The difference between conductors and insulators was also established, while Coulomb, Priestley and Cavendish proved that electric and magnetic forces diminished as the square of the distance. This enabled Gauss to include them with gravitation in a general mathematical treatment of the inverse square law. To Gauss we also owe a scientific system of electric units, for instance: a unit electric charge repels an equal similar charge at unit distance (one centimetre) in air with unit force (one dyne). If air be replaced by another medium, the electric or magnetic forces will be diminished in a ratio called the dielectric constant or the magnetic permeability.

In 1800 Volta of Pavia, extending an experiment of Galvani, found that a series of little cells containing brine or dilute acid, in which were dipped plates of zinc and copper, gave a current, proved by Wollaston to be electric, with tension much less than that of frictional electricity, and quantity much more. Nicholson and Carlisle showed that the current evolved hydrogen and oxygen when passed through water—new and striking evidence for the compound nature of that substance, and the first experiment in electro-chemistry. The connexion thus revealed between electricity and chemistry caused much speculation. Berzelius regarded every compound as formed by the union of two oppositely electrified parts. In 1807 Sir Humphry Davy (1778-1829) decomposed soda and potash, thought to be elements, and separated the remarkable metals sodium and potassium. The products of decomposition appear only at the terminals, and Grotthus and Clausius explained that fact by imagining a free exchange of partners along lines through the liquid, the opposite atoms at the ends of the chain being set free.

The next great advance was made by Michael Faraday (1791-1867), who had been Davy's assistant in the laboratory of the Royal Institution. Faraday, in 1833, with Whewell's advice, introduced a new terminology which we still use—electrolysis, ions, anode, cathode, etc. He reduced the complexity of the subject to two simple statements: the mass liberated is proportional (1) to the strength of the current and to the time it flows, that is, to the total amount of electricity passing through the liquid; and (2) to the chemical equivalent weight. For instance, when one ampere flows for a second through an acid solution, 1.044×10^{-5} gram of hydrogen is liberated, while from the solution of a silver salt 0.001118 gram of silver is deposited. The weight of silver is easy to measure and may be used as a practical definition of the ampere.

Faraday's work shows that we must regard the flow of electricity through conducting liquids as the carrying of charges by moving atoms or ions—positive charges in one direction and negative in the other. At a later date Helmholtz pointed out that it follows that electricity, like matter, is atomic, the fundamental unit being the charge on one monovalent ion.

Other properties of currents soon discovered were the heating effect, now used in lighting and warming our houses, and the deflexion of a magnetic needle, described by Oersted in 1820. This result was extended by Ampère, who showed that currents also exert forces on each other, while a circular coil gives a magnetic force at its centre equal to $2\pi c/r$, a result which enables us to measure a current in absolute C.G.S. units. Ampère's work led to the invention of a practical electric telegraph.

About 1827 G. S. Ohm (1787–1854) replaced the prevalent vague ideas of 'quantity' and 'tension' by definite conceptions of current strength and electromotive force. He showed that the current was proportional to the E.M.F., that is $c = yE = \dfrac{E}{R}$, where y is a constant known as the conductivity, and R its reciprocal the resistance, a relation known as Ohm's Law.

Electro-magnetism. In 1831 Faraday discovered that, when an electric current is started or stopped in a coil of wire, a small transient current is produced in another neighbouring coil. He also rotated a copper disc between the poles of a magnet, and obtained a steady electric current by touching the axle and periphery with wires from a galvanometer. This was the first dynamo, afterwards the basis of electrical engineering.

Faraday realized the importance of the medium across which electric and magnetic forces stretch; he imagined lines of force, or chains of particles in 'dielectric polarization'. His ideas were put into mathematical form from 1864 onwards by Clerk Maxwell. As a change in 'dielectric polarization' spreads through a medium, it travels as an electromagnetic wave, which, Maxwell showed, moved with a velocity $v = 1/\sqrt{\mu\kappa}$, where μ and κ are the magnetic permeability and dielectric constant of the medium. Since the electric and magnetic forces depend on μ and κ, the units derived from them also so depend, and by comparing the electro-static to the electro-magnetic units the value of v can be found. It proved to be about 3×10^{10} centimetres, or 186,000 miles, a second, equal to the velocity of light. Maxwell concluded that light is an electro-magnetic wave, so that one aether will carry both. At the time this strengthened belief in the aether, though it was uncertain whether electric waves were mechanical vibrations in a quasi-rigid solid, or light electro-magnetic oscillations, the meaning of which remained unknown. Electro-magnetic waves were first demonstrated experimentally by Hertz in 1887, but the aether, if there be an aether, is now crowded with 'wireless waves'—a consequence of Maxwell's equations—and, however they are conveyed, it is certainly not 'on the air'.

The Atomic Theory. From the days of Democritus onward the atomic idea in a vague form appeared at intervals. Boyle and Newton used it in their physical speculations, but, at the beginning of the nineteenth century, Dalton made it into a definite quantitative theory, based on the numerical facts of chemical combination.

Lavoisier and others had proved that a chemical compound however obtained is always made up of the same amount of the same constituent parts—water, for instance, always consists of one part of hydrogen combined with eight of oxygen, giving 8 as the 'combining weight' of oxygen. Nevertheless, some chemists, among them Bertholet, did not believe in this constancy. But John Dalton (1766–1844) saw that the properties of gases were best explained by atoms, and then pointed out that, on that theory, the combining weights in chemical combination gave the relative weights of the atoms also. He drew up a list of twenty such atomic weights. This formulation was too simple; Dalton thought that, if only one compound of two elements was known, it was fair to assume it was formed atom for atom. This is not always true, indeed it put Dalton wrong about water.

Gay-Lussac showed that gases combine in volumes that bear simple ratios to each other, and Avogadro pointed out that this must apply also to the numbers of their combining atoms. In 1858 Cannizzaro saw that it was necessary to distinguish between the chemical atom, the smallest part of matter which can enter into chemical action, and the physical molecule, the smallest particle which can exist in the free state. The simplest form of Avogadro's hypothesis is to suppose that equal volumes of gases at the same temperature and pressure contain the same number of molecules, a result which follows also from the kinetic theory of gases. In the formation of water from its elements we have two volumes of hydrogen combining with one of oxygen to form two of water vapour. The simplest theory which will explain these facts is to suppose that each molecule of hydrogen or oxygen contains two atoms, and that

$$2H_2 + O_2 = 2H_2O,$$

two atoms of hydrogen combining with one of oxygen to form one molecule of water. Oxygen is therefore said to be a divalent element. The concept of valency has been much used in chemical theory. From the considerations given above it follows that the atomic weight of oxygen is not 8 but 16. Thus Dalton's combining weights need to be considered in the light of other experiments before we can assign to the elements their true atomic weights. This was first done systematically by Cannizzaro.

The number of elements known has grown from Dalton's 20 to 92, each new method of research disclosing elements previously unknown. The results of spectrum analysis and electric currents have already been described. In 1895 the third Lord Rayleigh observed that nitrogen separated from the air had a density slightly greater than that of nitrogen extracted from its compounds, and he and Sir William Ramsay traced the difference to the presence in air of a heavier, chemically inert gas which they named 'argon'. Four other inert gases—helium, krypton, neon and xenon—were soon afterwards discovered, and, with argon, placed as a group of zero valency.

A connexion between atomic weights and physical and chemical properties was sought by several chemists, the most successful being the Russian Mendeléeff (1834–1907). On arranging the elements in order of ascending atomic weights, a periodicity appeared, each eighth element having somewhat similar properties. The table gave a means of assigning correct atomic weights to elements of doubtful valency. Blanks in the table were filled hypothetically by Mendeléeff, who thus predicted the existence and properties of unknown elements, some of which were afterwards discovered. As measured chemically, many elements (but not all) have atomic weights approaching whole numbers. This suggested to Prout and others that all elements were composed of hydrogen, or at all events of some common basis. But this idea was beyond the theoretical and experimental powers of the time to test, and remained for a later age to verify.

Organic Chemistry. The bodies of plants and animals are composed of complex chemical substances for the most part based on the remarkable element carbon, the atoms of which can combine with each other. For long it was thought that these substances could only be formed by vital action, but the artificial preparation of urea by Wöhler in 1828 showed that one such substance at all events could be made in the laboratory. Other artificial productions followed, and Emil Fischer in 1887 built up fructose (fruit sugar) and glucose (grape sugar) from their elements.

The method of determining the percentage composition is due to Lavoisier, Berzelius, Liebig and others. The compound is burned in the oxygen from copper oxide and the products of combustion, such as water and carbon dioxide, weighed. In this way the composition of innumerable organic bodies has been determined. One surprising result was the discovery that certain compounds, quite different in physical and chemical properties, had the same percentage composition—for example, urea and ammonium cyanate. Berzelius (1779–1848) explained this isomerism as due to differences in the connexions between the atoms in the molecule. The same phenomenon is seen in charcoal and the diamond; they both consist of carbon.

Thus empirical formulae, such as C_2H_6O for alcohol, became constitutional formulae like

$$\begin{array}{c} \text{H} \quad \text{H} \\ | \quad | \\ \text{H}-\text{C}-\text{C}-\text{OH} \\ | \quad | \\ \text{H} \quad \text{H} \end{array}$$

for the same substance. In 1865 Kekulé showed that the properties of benzene, C_6H_6, indicated a closed ring of the six carbon atoms. If one or more of the hydrogen atoms are replaced by other atoms or groups of atoms, the more complex aromatic compounds can be represented. By these structural formulae, organic chemistry has been rationalized; from the possibilities suggested by the formulae, new compounds have been predicted and isolated.

The chemistry of coal-tar has developed into an enormous industry. Unverdorben and later Hofmann isolated from tar a substance to which the name of aniline was given, and in 1856 W. H. Perkin (Senior) obtained aniline purple or mauve, the first aniline dye, soon followed by countless others, especially in Germany. In 1878 E. and

O. Fischer studied their constitution, and found the basis of many of them in triphenylmethane. About 1897 indigotin made from phenylglycine began to drive natural indigo off the market. Synthetic organic drugs appeared with antipyrene (1883), phenacetin (1887) and acetyl salicylic acid or aspirin (1899). Then came Ehrlich's salvarsan and other specific remedies described later.

In 1844 Mitscherlich, who had previously pointed out the connexion between atomic constitution and crystalline form, showed that isomers of tartaric acid, though possessing the same structural formulae, had different optical properties. In 1848 Pasteur (1822–1895), when recrystallizing racemates, got two varieties of crystal, related to each other as a right hand to a left hand or an object to its image. When the two kinds of crystals were picked out, separated and redissolved, one solution was found to rotate the plane of polarization of light to the right and the other to the left. Many of the substances obtained from living bodies are optically active in this way, while, if they are synthesized in the laboratory, they are inactive.

In 1863 Wislicenus and in 1874 Le Bel and Van't Hoff inferred that the atoms in the molecules of the two varieties must be arranged differently in space, giving formulae related to each other as object and image. They pictured a carbon atom C at the centre of a tetrahedron, with four or more different atoms or group of atoms linked to it—what is called an asymmetric carbon atom. Similar phenomena have been found with a few other elements, especially nitrogen.

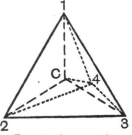

Fig. 12. Asymmetric carbon atom

A group of atoms, or radical, such for instance as hydroxyl OH, may hold together through a series of reactions. This suggested a theory of types, according to which chemical compounds were once classified.

Gradually an enormous number of organic compounds have been isolated, and many synthesized from their elements. Liebig grouped them as members or derivatives of one or other of three great classes:

(1) Proteins, containing carbon, hydrogen, nitrogen, oxygen, and sometimes sulphur and phosphorus.

(2) Fats, containing carbon, hydrogen and oxygen.

(3) Carbohydrates, containing carbon, hydrogen and oxygen, the hydrogen and oxygen being present in the proportions in which they form water.

Of these compounds the proteins are the most complex. They break down into constituents known as amino-acids. In 1883 Curtius built up a substance which gave a protein-like reaction. Fischer examined it, and devised several methods of combining amino-acids into bodies resembling the peptones which are formed by the action of digestive ferments on proteins. Thus, before the end of the century, progress had been made towards determining the nature and possible methods of synthesis of some constituents of living organisms. Moreover, the knowledge gained in rationalizing organic chemistry made clearer many problems in other branches of chemical science.

Chemical Action. The causes and mechanism of chemical affinity and chemical action have been discussed from early times, and Newton gave much attention to the subject. In 1850 Wilhelmy measured the rate of 'inversion' of cane sugar in presence of an acid —the dissociation of molecules of sucrose into the simpler ones of dextrose and laevulose. He found that, as the concentration of the cane sugar got less, the amount changing also diminished in a geometrical progression. This means that the number of molecules dissociating is proportional to the number present at any instant— as indeed we should expect if the molecules dissociated independently of each other in what is called a monomolecular reaction.

But if two molecules react with each other—a dimolecular reaction —the rate of change must depend on the frequency of collision, a frequency proportional to the product of the concentrations or active masses of the two reagents; if the molecular concentrations are equal, this product will be the square of the concentration.

As Bertholet discovered, some actions are reversible, able to go in either direction. If two compounds AB and CD are forming AD and CB, and the reverse change is also going on, equilibrium must

THE NINETEENTH CENTURY

clearly be reached when the rates of the opposite reactions are equal. The process is written as

$$AB + CD \rightleftharpoons AD + CB.$$

This dynamical equilibrium was first formulated by A. W. Williamson in 1850. A fuller statement of the mass law was given by Guldberg and Waage in 1864 and by Van't Hoff in 1877.

The inversion of cane sugar goes much faster in presence of an acid. The acid facilitates the action but is itself unchanged, and was called by Berzelius a 'catalyst'; similarly, finely divided platinum brings about the combination of hydrogen and oxygen. Many such actions are known both in chemistry and in physiological processes, where the catalysts are called enzymes or hormones. Catalysts are also now frequently used in industrial chemistry.

Solution. The solution of substances in water and other liquids is familiar. Sugar dissolves freely, while metals are insoluble. Air is only slightly soluble in water, but ammonia and hydric chloride are very soluble. Some solutions, such as most of those of salts and mineral acids, conduct electricity well, while pure water and solutions of sugar have very low conductivity.

Faraday's work proved that electrolytic conduction is a motion in opposite directions of positively and negatively electrified ions, every ion of the same chemical valency carrying the same charge. In 1859 Hittorf saw that the unequal dilution of the solution round the two electrodes gave a means of comparing the velocities of the opposite ions, since the faster moving ion must concentrate more salt round the terminal towards which it moves. In 1879 Kolrausch measured the conductivity of solutions, using alternating currents and a telephone instead of a galvanometer, to avoid polarization at the electrodes. He found that Ohm's law holds good, so that the smallest electromotive force produces a current; there must therefore be freedom of interchange in the body of the liquid. He also pointed out that the conductivity gave a means of finding the sum of the opposite ionic velocities, which, combined with Hittorf's values for their ratios, gave the individual velocity of each ion. For a potential gradient of one volt per centimetre, hydrogen moves through water with a velocity of about 0·003 centimetre per second, while the ions of salts range round 0·0006. Oliver Lodge verified the speed of hydrogen, and I measured that of some other ions by watching their motion if coloured or by the formation of precipitates.

Pressures are set up in vegetable cells by the passage of water through the containing membranes, and Pfeffer measured this so-called osmotic pressure for solutions of cane sugar, using artificial membranes deposited chemically in the walls of porous pots. Van't Hoff pointed out that Pfeffer's results showed that osmotic pressure, like gas pressure, varies inversely as the volume, and has about the same value as has an equal molecular concentration of a gas. The existence of the pressure makes it possible to imagine the osmotic cell acting as the cylinder of Carnot's ideal engine, and Van't Hoff was thus able to apply thermodynamic reasoning to solutions, opening a new field of research. In dilute solutions the osmotic pressure, both theoretically and experimentally, has the gas value, but not in strong solutions, investigated by the Earl of Berkeley and E. G. J. Hartley. Van't Hoff connected osmotic pressure with other properties of solutions, such as the depression of the freezing point, more easily measured. These relations with gases do not prove that osmotic pressure is due to the same cause as gas pressure; thermodynamic reasoning gives results, but has nothing to say about mechanism. The pressure may be due to impact, to chemical attraction or some other unknown cause.

In 1887 the Swede Arrhenius, knowing that the osmotic pressures of electrolytes were abnormally great, so that the pressure in a dilute solution of potassium chloride or other binary salt was about twice the value for sugar or for a gas, accepted the conclusion that this showed the number of pressure-producing particles to be greater than in non-electrolytes. He therefore put forward the theory that electrolytes are dissociated into ions in solution. For instance, with potassium chloride there are some neutral molecules of KCl, and some dissociated ions K^+ and Cl^-. As the solution is diluted, there is more dissociation, till all the salt is resolved into its ions, osmotic pressure and chemical activity per unit mass increasing *pari passu*.

THE NINETEENTH CENTURY

The ions are probably combined with the solvent, perhaps carrying an atmosphere of the liquid with them as they move.

On these lines, thermodynamics, especially as developed by Willard Gibbs (p. 99), and electrical science have been combined in an ever-growing extension of theoretical knowledge and of practical industrial applications. Moreover, the theory of solution gave the idea of ions to those physicists who later on investigated the conduction of electricity through gases and revolutionized modern physical science.

Colloids. The distinction between compounds that will crystallize, named crystalloids, and those that form much larger solid structures, called colloids, was recognized by Thomas Graham in 1850, and explained by the size of the colloid particles. The solution of a crystalloid, sugar (say) or salt, is a homogeneous liquid, but a colloid forms a system of two phases, with a surface of separation between them. Some colloid particles can be seen in a microscope; for others the ultra-microscope is needed. In this the observer looks at right angles to a strong beam of light, which is scattered by the particles, so that, if they are not much smaller than the wave-length of the light used, they show as separate bright discs, kept in oscillatory (Brownian) movement by the collision of molecules.

The protoplasm which forms the contents of living cells consists of colloids, the nucleus being more solid than the remainder; hence the importance of colloids in physiology. They appear also in agricultural science, for soil is now known to be a complex living structure of organic and inorganic colloids, in which micro-organisms, bacteria and others, play an essential part in breaking down the raw material into substances fit for plant food. Again, the colloids in clay control its physical texture, which only becomes porous and fertile when the plastic clay is coagulated.

Colloid particles move in an electric field, and must therefore carry electric charges. Sir W. B. Hardy found that, when the surrounding liquid was slowly changed from acid to alkaline, the charges on certain colloids were reversed. At the 'iso-electric point', where the charge was neutralized, the system became unstable and the colloid was coagulated.

It was known to Faraday and to Graham that some colloid solutions are coagulated by salts. In 1895 Linder and Picton found that the average coagulative power of mono-valent, di-valent and tri-valent ions were in the ratios of 1 : 35 : 1023. In 1899 I investigated the problem on the theory of probability. The electric charge on an ion is proportional to its chemical valency, and so it will need the conjunction of two tri-valent ions, three di-valent or six mono-valent ions to bring the same charge near a colloid particle. Calculation shows that the coagulative powers should be as $1 : x : x^2$, where x is some unknown number depending on the structure of the system. Putting $x=32$, we get 1 : 32 : 1024 to compare with the observed values. This can only be an approximate result; it ignores disturbing factors. But the method explains the high effect of tri-valent ions, and similar principles of probability are applicable to chemical reaction itself.

CHAPTER VIII. *NINETEENTH-CENTURY BIOLOGY*

Biology and its Effects. In the seventeenth and eighteenth centuries it was astronomy that affected most profoundly the minds both of philosophers and of ordinary men. Copernicus dethroned the Earth from its ancient position as the centre of the Universe, while Galileo and Newton proved that the heavenly bodies, no longer divine and incorruptible, move in accordance with terrestrial dynamics. Man's outlook on the cosmos was revolutionized.

In the nineteenth century this change had been assimilated and no longer caused distress. Physicists banished philosophy from their laboratories, and did their work by the light of a common-sense realism, never doubting that their discoveries showed the actual structure of the world. Mach pointed out that science does but create a model of what our senses tell us about nature, but few men of science listened.

The next revolution in scientific and philosophic thought came from biology, with Darwin as its chief figure. The old theory of evolution was made credible by his concept of natural selection, and man was forced to recognize his true place in the animal kingdom. Then evolutionary ideas spread from biology to other departments of knowledge.

Physiology. In the second half of the eighteenth century, the difficulty of explaining physiological processes led to the almost universal adoption of vitalism, the theory that living matter is above the range of physics and chemistry. But by the middle of the nineteenth century success in applying physical and chemical methods to biology, aided by the invention of physiological apparatus, swung opinion in the other direction. For most studies in physiology it seems necessary to accept physics and chemistry at all events as guides to the elucidation of details. The problem of organisms as wholes is different and more complex.

In the early years of the nineteenth century, Johannes Müller collected all available physiological knowledge in his *Outlines of*

Physiology and himself worked on nervous action. He proved that sensation depends on the nature of the sense organ and not on the mode of stimulation; light, pressure or mechanical irritation acting through the optic nerve on the retina, all produce luminous sensations, thus confirming the philosophic view that man's unaided senses give him no real knowledge of the external world.

During 1833 the American army surgeon Beaumont published facts about digestion observed in a patient with a gun-shot wound which left a hole into his stomach. Claude Bernard (1813–1878) studied the same condition in animals, and showed that pancreatic juice disintegrates the fats discharged by the stomach into the duodenum, decomposes them into fatty acids and glycerine, converts starch into sugar and dissolves proteins.

Dumas and Boussingault taught that, while plants absorb inorganic bodies and build them up into organic substances, animals, essentially parasitic, live by breaking down these substances into simpler compounds. But Bernard proved by experiments on dogs that the liver formed the carbohydrate dextrose from the blood, and also discovered that the liver produced a starch-like substance, named by him glycogen, which gave dextrose by a process of fermentation. Thus he showed that animals could build up some organic substances, though that function is generally performed by plants.

Bernard also explained the poisonous action of carbon monoxide gas by showing that it irreversibly displaces oxygen from the haemoglobin in the red blood corpuscles; the haemoglobin then becomes inert and can no longer carry oxygen to the tissues of the body. Bernard held that the function of the vital mechanism was to keep constant the conditions of the internal environment.

Schleiden in plants and later Schwann in animals established the theory that their tissues are made up of vast numbers of separate cells, tracing them to their origins in the embryos. Harvey and Wolff had put embryology on a sound basis, and it was taken up again by von Baer (1792–1876), who traced the multiplication and differentiation of cells in embryonic development. He rediscovered the ovum in mammals, and did much to create modern embryology. He opposed a prevalent theory that the history of the individual recapitulates the history of the species.

NINETEENTH-CENTURY BIOLOGY

Von Mohl examined the contents of cells and called the plastic substance within the cell wall by the name of protoplasm. Schultz described the cell as a mass of nucleated protoplasm, the physical basis of life. Virchow carried the cell theory into the study of diseased tissues, and discovered that the white corpuscles in the blood can engulf and destroy poisonous bacteria.

The general principle of the conservation of energy requires that the physical activities of the body must be maintained by the chemical and thermal energy of the food taken in. Atwater and Bryant, by experiments in America, gave 4·0 calories as the heat value of one gram of protein or carbohydrate and 8·9 for fats. T. B. Wood, working on farm animals, separated the maintenance ration needed to keep the animal in a stationary state from the additional food required for growth, milk-production, etc. A man doing no muscular work wants food to the equivalent of about 2450 calories a day, and one doing heavy labour 5500.[1] The energy of the food taken in has been found to be equal to the output in muscular work, heat and excrement, in full accord with the principle of the conservation of energy.

During the period now under review, much advance was made in the study of the nervous system. The localization of the functions of the brain had often been suggested, and was investigated by Gall, who taught that the grey matter was the active instrument of the nervous system and the white matter the connecting links. Majendie proved an idea due to Bell, that the anterior and posterior roots of spinal nerves had different functions. With Marshall Hall they established the difference between volitional and unconscious or reflex action. Such common acts of life as breathing, sneezing, etc. may be taken as reflexes, and other processes, formerly thought to involve complicated mental operations, were held to be reflexes, especially by Charcot later in the century. Bernard investigated the function of the vaso-motor nerves, showing that they are put into action involuntarily by sensory impulses and control the blood

[1] All these figures are expressed in the so-called 'great calory', 1000 times the unit used in physics, which is the amount of heat needed to raise one gram of water through one degree Centigrade.

vessels. E. H. and E. F. Weber discovered inhibiting actions, such as the stoppage of the beat of the heart by stimulation of the vagus nerve.

The study of catalytic actions, as found in inorganic chemistry, was extended to many processes going on in living organisms. In 1878 Kühne, who did much to trace their action, gave to these organic catalysts the name of enzymes (ἐν ζύμῃ, in yeast). Like other catalysts, enzymes, without themselves being altered, facilitate reactions in either direction, as oil helps a machine. Their effect is often specific—one reaction, one enzyme. They are often colloids and may carry electrical charges, indeed ions may act as catalysts, as in the inversion of cane sugar in presence of acids. Among the more important enzymes are amylase, which decomposes starch, pepsin which breaks up proteins in an acid medium as trypsin in an alkaline, lipase which decomposes esters and so on. We shall find later that this subject is of growing importance, especially in the study of glands and their secretions. As early as 1884 Schiff found that the effects of the removal of the thyroid gland from an animal could be overcome if the animal were fed with an extract from the gland. This result was soon applied to the variety of human idiocy known as cretinism, which was found to be due to a failure in the thyroid gland, and many children who in former times would have been hopeless imbeciles for life have been made into intelligent and happy beings.

Bacteriology. About 1838 Cagniard de Latour and also Schwann discovered that the yeast used in fermentation consists of minute living cells. Schwann also proved that putrefaction was a similar process, and showed that neither fermentation nor putrefaction would occur if the substance were heated to destroy living organisms and afterwards protected from air save that which had passed through red-hot tubes.

Louis Pasteur (1822–1895) confirmed these results, and extended them to the souring of beer and wine, the silk-worm disease, and (the greatest benefit of all) to rabies in animals and hydrophobia in mankind. As Jenner produced immunity from small-pox by vaccination, so Pasteur found by experiment on dogs that he could give

NINETEENTH-CENTURY BIOLOGY

them immunity from rabies by inoculation with attenuated poison, and thus found a cure for men who had been bitten. Each of these affections or diseases is due to a specific microbe or bacterium, though some enzymes expressed from them produce the same effect.

Lister, applying Pasteur's discoveries to surgery, used first phenol (carbolic acid) as an antiseptic, and then found an aseptic treatment in careful cleanliness. Lister's methods, together with the discovery of anaesthetics by Davy, Morton and Simpson, made safe many surgical operations previously impossible.

Koch found that the spores of the anthrax bacilli were more resistant than the bacilli themselves, an observation important in the technique of bacteriology. He also discovered the micro-organism responsible for tuberculosis. The life-history of some pathogenic organisms is complex; they may need several kinds of host. This is true of malaria, carried by mosquitoes, and of Maltese fever, the microbe of which passes part of its life in goats.

We shall describe later the many discoveries of ultra-microscopic viruses; but one such, responsible for tobacco disease, was investigated by Ivanovski in 1892, and another by Löffler and Frosch in 1893. The latter found that the infection of foot-and-mouth disease would pass freely through a filter which stopped ordinary bacteria, and would still affect animals in series. The nature of these viruses is still uncertain.

It is clear that essential parts of our knowledge of the organs and functions of the human body, with all their beneficial effects on medicine and surgery, could only have been reached by experiments on animals, carried on from early days till the present. Those who try to stop further advance by this method take on themselves a fearful moral responsibility, not lightened by their frequent ignorance of the facts.

The Carbon and Nitrogen Cycles. The relations between plants and animals were studied further by several investigators, leading up to the work of Liebig. Plants are built up by the action in sunlight of the green colouring matter chlorophyll, which uses the Sun's energy to decompose the carbon dioxide in the air, liberating oxygen and combining the carbon in the complex organic molecules of plant tissues. In the absorption spectrum of chlorophyll, the maximum absorption coincides with the maximum energy of the solar spectrum—a remarkable adaptation, however produced, of means to ends.

Animals live either on plants or on each other, and so all are ultimately dependent on the energy of the Sun. In breathing they oxidize carbon compounds into carbon dioxide and the derivatives they need; other substances are excreted, and the residual energy appears as bodily heat. Plants also slowly give out carbon dioxide, though in sunlight this is masked by the reverse process. Thus both animals, and to a lesser degree plants, return to the air the carbon dioxide which plants have removed. The waste organic solid products are deposited in the ground, where they are attacked by teeming bacteria and are broken down into harmless (indeed useful) inorganic bodies, while more carbon dioxide is poured into the air. Thus the carbon cycle is completed.

The corresponding cycle for nitrogen was worked out at a later date. In his *Georgics*, Virgil recommends the farmer to grow beans, vetch or lupins before wheat, but the reason for the success of this sound advice was only discovered in 1888 by Hellriegel and Wilfarth. The roots of leguminous plants produce nodules containing bacteria which can fix nitrogen from the atmosphere, convert it into protein and pass it on to the plant. Moreover, Vinogradsky found in the soil bacteria which obtain nitrogen direct from the air. Waste nitrogenous products are converted in the soil, again with the help of bacteria, into ammonium salts and finally into nitrates, the best form of nitrogen for plant food. Soil is a live physical, chemical and biological structure which needs humus as well as mineral salts to keep its balance.

Liebig demonstrated the importance of mineral salts in agriculture, but he overlooked the need for nitrogen. This was discovered by Boussingault, and by Gilbert and Lawes of Rothamsted, who introduced artificial manures to the farmer.

Geography and Geology. In 1784 the Ordnance Department measured a baseline on Hounslow Heath and began trigonometrical

NINETEENTH-CENTURY BIOLOGY

surveys, which enabled d'Anville and others to make accurate maps. Von Humboldt studied the variations of the Earth's magnetic force and the magnetic storms he was the first to observe.

Reviving the practice of sending out exploratory expeditions, in 1831 the *Beagle* was despatched for scientific observations round Patagonia and Tierra del Fuego, with Charles Darwin on board as official naturalist. A few years later Joseph Hooker joined Sir James Ross in his Antarctic voyage, and in 1846 T. H. Huxley sailed as surgeon on the *Rattlesnake* and spent several years surveying and charting in Australian waters. Thus three men, who had much to do with the coming and establishment of the theory of evolution, served an apprenticeship on voyages of scientific exploration. Finally in 1872 the *Challenger* was sent to cruise for some years in the Atlantic and Pacific to take records of oceanography, meteorology and natural history. Maury of the United States navy took up the problems of winds and currents as they were left a century and a half earlier by Dampier, and improved the navigation of ocean routes. A study of the life of the sea revealed countless forms, from the drifting microscopic matter named 'plankton' by Henson to the fish that follow it for food.

Hutton's uniformitarian theory was much strengthened when Sir Charles Lyell collected in his *Principles of Geology* (1830–1833) all available evidence about the formation of strata and the origin of fossils. Fossils indicate changes of climate at definite times, in this supported by the evidence of glaciation found by Agassiz and Buckland about 1840. Flint implements and carved pieces of bone and ivory enabled Lyell in 1863 to place man in his position in the series of organic types, and to show that his existence on the Earth must have extended over much longer ages than those contemplated by the Biblical Chronology current at the time. Much evidence has accumulated since Lyell's day, and it now seems likely that men appeared on the Earth somewhere between one and ten million years ago.[1]

Evolution and Natural Selection. The idea of evolution was known to some of the Greek philosophers. The Atomists seem to have

---

[1] See Chapter I, p. 1.

thought that each species arose independently, but, in their belief that only those types survived which were fitted to the environment, they touched the concept of natural selection. By the time of Aristotle, speculation had suggested that more perfect types had not only followed less perfect ones but actually had developed from them. But all this was guessing; no real evidence was forthcoming.

When, in modern times, the idea of evolution was revived, it appeared in the writings of the philosophers—Bacon, Descartes, Leibniz and Kant—though some of them, Hegel for instance, took evolution in an ideal sense. But Herbert Spencer was preaching a full evolutionary doctrine in the years just before Darwin's book was published, while most naturalists would have none of it. Nevertheless, a few biologists ran counter to the prevailing view, and pointed to such facts as the essential unity of structure in all warm-blooded animals.

The first complete theory was that of Lamarck (1744–1829), who thought that modifications due to the environment, if constant and lasting, would be inherited and produce a new type. Though no evidence for such inheritance was available, the theory gave a working hypothesis for naturalists to use, and many of the social and philanthropic efforts of the nineteenth century were framed on the tacit assumption that acquired improvements would be inherited. Similar views were advocated by Saint-Hilaire and by Robert Chambers, whose book *Vestiges of Creation* helped to prepare men for the acceptance of evolution.

But the man whose book gave both Darwin and Wallace the clue was the Rev. Robert Malthus (1766–1834), sometime curate of Albury in Surrey. The English people were increasing rapidly, and Malthus argued that the human race tends to outrun its means of subsistence unless the redundant individuals are eliminated.

This may not always be true, but Darwin writes:

In October 1838 I happened to read for amusement Malthus on Population, and being well prepared to appreciate the struggle for existence which everywhere goes on from long continued observation of the habits of animals and plants, it at once struck me that under these circumstances favourable variations would tend to be preserved,

NINETEENTH-CENTURY BIOLOGY

and unfavourable ones to be destroyed. The result of this would be the formation of new species. Here then I had a theory by which to work.

Darwin then spent twenty years collecting countless facts and making experiments on breeding and variation in plants and animals. By 1844 he had convinced himself that species are not immutable, but worked on to get further evidence. On 18 June 1858 he received from Alfred Russel Wallace a paper written in Ternate in the space of three days after reading Malthus' book. Darwin saw at once that Wallace had hit upon the essence of his own theory, and placed himself in the hands of Lyell and Hooker, who arranged with the Linnaean Society to read on 1 July 1858 Wallace's paper together with a letter from Darwin to Asa Gray dated 1857 and an abstract of his theory written in 1844. Then Darwin wrote out an account of his labours, and on 24 November 1859 published his great book *The Origin of Species*.

Charles Robert Darwin (1809–1882) [see Plate VI, facing p. 122] was the son of an able doctor, Robert Waring Darwin of Shrewsbury, and grandson of Erasmus Darwin, who had ideas about evolution, and of Josiah Wedgwood, the potter of Etruria, who also showed the scientific ability which has now appeared in five generations of Darwins.

In any race of plants or animals, the individuals differ from each other in innate qualities. Darwin offered no explanation of these variations, but merely accepted their existence. When the pressure of numbers or the competition for mates is great, any variation in structure which is of use in the struggle has 'survival value', and gives its possessor an improved chance of prolonging life and leaving offspring. That variation therefore tends to spread through the race by the elimination of those who do not possess it, and a new variety or even species may be established. As Huxley said, this idea was wholly unknown till 1858.

Huxley, Asa Gray, Lubbock and Carpenter accepted Darwin's theory at once, and Lyell was convinced by 1864. Huxley called himself 'Darwin's bulldog', and said the book was like a flash of lightning in the darkness. He wrote:

It did the immense service of freeing us from the dilemma—Refuse to accept the Creation hypothesis, and what have you to propose that can be accepted by any cautious reasoner? In 1857 I had no answer ready, and I do not think that anyone else had. A year later we reproached ourselves with dulness for being perplexed with such an enquiry. My reflection when I first made myself master of the central idea of the *Origin* was 'How extremely stupid not to have thought of that!'

The hypothesis of natural selection may not be a complete explanation, but it led to a greater thing than itself—an acceptance of the theory of organic evolution, which the years have but confirmed. Yet at first some naturalists joined the opposition, and the famous anatomist Sir Richard Owen wrote an adverse review. To the many, who were unable to judge the biological evidence, the effect of the theory of evolution seemed incredible as well as devastating, to run counter to common sense and to overwhelm all philosophic and religious landmarks. Even educated man, choosing between the Book of Genesis and the *Origin of Species*, proclaimed with Disraeli that he was 'on the side of the Angels'.

Darwin himself took a modest view. While thinking that natural selection was the chief cause of evolution, he did not exclude Lamarck's idea that characters acquired by long use or disuse might be inherited, though no evidence seemed to be forthcoming. But about 1890 Weismann drew a sharp distinction between the body (or soma) and the germ cells which it contains. Somatic cells can only reproduce cells like themselves, but germ cells give rise not only to the germ cells of a new individual but to all the many types of cell in his body. Germ cells descend from germ cells in a pure line of germ plasm, but somatic cells trace their origin to germ cells; from this point of view, the body of each individual is an unimportant by-product of his parents' germ cells. The body dies, leaving no offspring, but the germ plasm shows an unbroken continuity. The products of the germ cells are not likely to be affected by changes in the body; so Weismann's doctrine offered an explanation of the apparent non-inheritance of acquired characters. It seems that, in spite of philosophers and philanthropists, nature is more than nurture and heredity than environment.

Plate VI   CHARLES ROBERT DARWIN

Plate VII   LORD RUTHERFORD OF NELSON

The supporters of pure Darwinism came to regard the minute variations as enough to explain natural selection and natural selection enough to explain evolution. But animal breeders and horticulturalists knew that sudden large mutations occur, especially after crossing, and that new varieties might be established at once. Then in 1900 forgotten work by Mendel was rediscovered and a new chapter opened.

Anthropology. Darwin's work started the study of man on a novel course, the idea of evolution underlying it all. Huxley's study of human skulls was inspired by the Darwinian controversy, and began the exact measurement of physical characters. The explorers who brought back plants and animals often brought also the industrial and artistic products of other peoples and the ceremonial objects of other religions or descriptions thereof. There was much material ready waiting for the anthropologist.

Since Lyell described what was known in his day of man in the geological record, many discoveries have been made. In the nineteenth century pictures of bison, prehistoric mammoths and other animals were found on the walls of caves. At Neanderthal in 1856 and at Spy in 1886 relics of primitive types of men appeared, and in 1893 Dubois discovered in the late Pliocene deposits in Java, bones which some think are those of a being intermediate in structure between anthropoid apes and early men. Man cannot be descended from any form of ape now existing, but he is at least a distant cousin of some of them.

Statistical methods were applied to men in the seventeenth century by a study of the Bills of Mortality, especially by Sir William Petty and John Graunt. The subject was again advanced by the Belgian astronomer Quetelet in 1835 and later. He found that the chest measurements of Scottish soldiers or the heights of French conscripts varied round the average as the bullets round the centre of a target, or the runs of luck at a gaming table. Graphically, as in the diagram, the curve, except that it is symmetrical on both sides, resembles that showing the velocities of molecules in a gas (p. 96). Similar considerations apply to such economic activities as insurance.

In 1869 Darwin's cousin Francis Galton applied these principles to mental qualities. Tracing the distribution of marks in an examination, he found the same laws as for physical qualities or molecular velocities. Most men have mediocre intellectual powers, and, as we pass towards genius at one end and idiocy at the other, the numbers fall off as the curve shows. A Senior Wrangler obtained on the average about thirty times the number of marks of the lowest honours man, while the pass men, had they taken the same examination, would

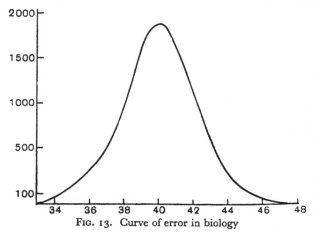

FIG. 13. Curve of error in biology

presumably have got still fewer marks. The differences in ability are clearly enormous, and the democratic idea that men are born equal is demonstrably false.

By searching books of reference, Galton examined the inheritance of ability; for instance he found that the chance of the son of a judge showing great ability was about five hundred times as high as that of a man taken at random, and for the judge's father it was nearly as much. While no prediction can be made about individuals, on the average of large numbers the inheritance of ability is certain.

Thus we must give up the idea that the nation consists of a number of individuals of equal potential capacity, only waiting for education and opportunity, and look on it as an interwoven network of strains of various innate hereditary qualities, strains differing in character and value, and increasing or disappearing chiefly in accordance with natural or artificial selection. Almost any action, social, economic or legislative, will favour some of these strains at the expense of others, and alter the average biological character of the nation. This subject will be pursued in the light of more recent information in Chapter IX.

Nineteenth-Century Science and Philosophy. Till the days of Kant philosophers framed their systems in the light of physical science, but Hegel and the Hegelians, starting from *a priori* philosophy, constructed a theory of nature which seemed to men of science fantastic. In turn Hegel attacked physicists, and especially Newton as their exemplar. The poet Goethe, too, who had done good work in animal and vegetable anatomy, where the facts lay on the surface, failed when he touched physics. A flash of poetic insight assured him that white light must be simpler and purer than coloured, and Newton's theory wrong. He refused to consider facts disclosed by experiments and inferences drawn from them; the senses must reveal at once the truth about nature. Thus for a time the philosophers scorned the scientists and the scientists ignored the philosophers; thinking they were keeping clear of metaphysics, they accepted uncritically the structure of nature built by science as ultimate reality. Though the philosophy of science was studied in England by Boole, Jevons, Clifford and Spencer, they had little influence among scientists. Even when in 1883 Mach, treating mechanics from the historical point of view, and reviving the teaching of the philosophers from Locke to Kant, pointed out that science does but construct a model of what our senses tell us about nature, few listened. The separation between science and philosophy remained complete.

But soon in its turn science began again to influence philosophy. Lavoisier's proof that mass persisted through a series of chemical changes was brought into prominence to support the common-sense view that matter is real, and the striking success of physics and chemistry drew uncritical men to believe in matter and energy as ultimate realities.

This led in Germany to a revival of the materialism prevalent in

France a hundred years before. Some based their belief on physiology and psychology, but Büchner's book *Kraft und Stoff* showed that the idea of the reality of force and matter was an essential part of this philosophy. The establishment of the principle of the conservation of energy, when it came, was used to support the philosophic theory of mechanism and determinism.

William Thomson's proof that, in an isolated system, energy must continually become less available, and Clausius' entropy increase (p. 98), was extended, perhaps unwarrantably, to the cosmic problem, and used to predict a 'dead universe'—new evidence, some thought, for the spread of determinism and atheism. But, on the alternative hypothesis, man's soul, being immaterial and immortal, need not mind the dissolution of a physical universe.

Materialism and allied beliefs were most prevalent in Germany, but in some other continental countries ecclesiastical conservatism found means to suppress these views, till the struggle for political liberty was combined with that for intellectual freedom, and culminated in the revolutionary outbreaks of 1848. In the following years the industrial changes, which had already gone far in England, began to extend to the Continent. Science, especially chemistry, came to touch ordinary life more closely. In practical England, this process had little effect on religious orthodoxy, but in logical France and metaphysical Germany, it helped to swell the rising tide of mechanical and materialist philosophy.

And it was in Germany that Darwin's explanation of evolution by the principle of natural selection, when accepted by Haeckel and other biologists, went to build up a thoroughgoing *Darwinismus* and most strongly reinforced the materialist tendencies in both philosophy and political theory, the latter being used as a basis for communism especially by Marx and Engels. Thus natural selection ceased to be merely a tentative scientific theory, and became the basis of a philosophy of evolution. Meanwhile Darwin's own methods of careful observation and experiment fell into abeyance.

But, however much exaggeration was worked into a Darwinian philosophy, and however the inheritance of small variations and natural selection acting on them failed to explain the details of change of species, the revolution in thought due to the acceptance of the general principle of evolution was immense. Instead of a stationary world of living beings and a stationary human society, a dynamic picture of continual change emerged. The geological record showed a gradual change from simpler to more complex organisms, and from an ape-like precursor to *homo sapiens*. It was therefore perhaps natural to suppose that evolution should continue to mark progress both in natural history and in human morphology and human society; men ignored the fact that 'survival of the fittest' meant only fittest for the existing environment, and the possibility that adaptation to that environment might proceed in a downward direction. In morphology and sociology, evolution at first engendered a somewhat shallow optimism.

Psychology. The mind of man can be studied in two ways, rationally or empirically. If we accept some metaphysical system of the Universe, say that of the Roman Church or that of the German materialists, we can deduce rationally the place of the human mind in that system. On the other hand, making no assumptions, we can investigate mind by empirical observation, either by introspection with Locke, or by objective observation and experiment. This last method makes psychology a branch of natural science.

Early in the nineteenth century German universities were still combining rational psychology with cosmology and theology. In England and Scotland empirical psychology had appeared, and at first, led by Mill and Bain, followed the introspective method. In France mind was already being examined by external means as a physiological and pathological problem.

And about the middle of the century physical methods became general. Helmholtz studied physiological acoustics and the physiological basis of music and speech. E. H. Weber observed the limits of sensation, and found that the increase in stimulus necessary to cause an equal increase in sensation must rise in geometrical progression. Wundt made measurements on the sensation of time. Darwin studied the expression of emotion in animals and man.

Physical science is analytic, and may regard a problem successively from different aspects, mechanical, chemical or physiological, in each resolving the subject of study into simple concepts such as atoms or electrons. But biology, while using physical methods, also sees in each living being an organic whole, and each man feels a deep-seated consciousness of unity of being. Since each man's mind is fully accessible only to himself, this consciousness of unity cannot be investigated adequately by the methods of natural science; nevertheless the fact of organism is important.

In the human being physical and psychical phenomena clearly run parallel—they are simultaneous if not connected. The theory of psycho-physical parallelism, which can be traced back to Descartes, regards consciousness as an epi-phenomenon of the changes in the nervous system, which are more accessible to examination. Whether the two are distinct or connected and, if connected, whether the nervous system or the consciousness is the master, remains a problem perhaps insoluble. Again, is the feeling of unity a reflexion of a reality, and has the mind or soul an independent existence, or is it built up by the grouping together of sensations, perceptions and memories? Such are the questions posed, but not yet answered, by psychology.

CHAPTER IX. *RECENT BIOLOGY*

Genetics. Once more the rôles are reversed. During the last fifty years, biology, though adding greatly to knowledge, has followed along lines already laid down, while physics and chemistry, absorbing the system of Galileo, Newton and Dalton in wider generalizations, have revolutionized both science and philosophy.

For the most part, naturalists, accepting Darwin's work as final, had given up experimenting on variation, though some, like de Vries and William Bateson (1861–1926), were beginning to examine scientifically the large and sudden mutations familiar to practical horticulturalists. But in 1900 came the rediscovery of Mendel's forgotten writings, buried for forty years in the volumes of a local society.

G. J. Mendel (1822–1884), Abbot of the Königskloster at Brünn, made a series of experiments on the cross-breeding of the tall and dwarf varieties of green peas. All the hybrids were tall, outwardly resembling the tall parents, but when they were bred among themselves, on the average of large numbers three-quarters of the progeny were tall and one-quarter dwarf. The dwarfs in turn all bred true, while the others again produced pure dwarfs, pure talls and mixed talls. The facts are explained if we suppose that the germ cells of the original plants bear tallness or dwarfness as a pair of contrasted characters, and that tallness is 'dominant', so that it always shows if present, while dwarfness is 'recessive', and only shows if brought in from both sides. Then the probabilities of the conjunctions of different germ cells agree with the observed facts. Thus these biological qualities are reduced to indivisible units, and their behaviour to an exercise in the theory of probability. It is interesting to compare this result with the atomic and molecular theory and the recent quantum theory in physics. In neither case can we predict the fate of a single unit, but, among large numbers, the probable distribution can be foretold.

Many Mendelian characters have now been traced in plants and animals. They seldom show such simple relations as in green peas;

characters may be linked in pairs, so that one cannot appear without the other, or they may be incompatible and never be present together.

Investigations into cell structure by T. H. Morgan and others showed that within each cell-nucleus is a definite number of threadlike bodies which have been named chromosomes. If two germ cells unite, the fertilized ovum will contain double the original number of chromosomes. When the ovum divides, every chromosome divides likewise, the two parts going to the two daughter cells. Thus every body-cell contains a double set of chromosomes derived equally from the two parents. In the germ cells, at the last stage of transformation, the chromosomes unite in pairs, and the number is halved. The agreement of these cell phenomena with Mendelian inheritance was noticed by several people and it was put into definite form by Sutton.

Morgan and his colleagues in New York have worked out these relations more fully in the fruit-fly *Drosophila*, in which generations of large numbers succeed each other at intervals of ten days. They have found a numerical correspondence between the number of groups of hereditary qualities and the number of pairs of chromosomes, each being four. In the garden pea the number is seven, in wheat eight, in the mouse twenty, in man probably twenty-four. Even with twenty pairs of chromosomes, there will be over a million possible kinds of germ cells. It is easy to understand why no two individuals are identical.

The number of chromosomes in a reproductive cell is considered basic and is called the 'haploid' number. When fertilization occurs and two haploid numbers are brought together by the union of two nuclei, the resulting new individual is said to be 'diploid', but more than two haploid sets may appear. This polyploidy occurs in wheat, oats and cultivated fruits. Sweet cherries are diploids, plums hexaploids, while apples may be complex diploids or triploids. If a polyploid has an odd number of chromosomes which cannot be halved equally in the formation of reproductive cells, irregularities in chromosome distribution may take place, generally leading to sterility. Many varieties of fruit, Cox's Orange Pippin among apples, various plums, and all sweet cherries are unable to fertilize themselves, and need the near presence of some other variety to set their fruit.

Sex determination has been shown to depend on two factors, hereditary and developmental; the chromosomes which fix sex have, in some cases, been identified microscopically, while Crew has described the reversal of sex characters in fowls.

Again, genes are known which prevent the development of certain substances or qualities; some of these are fatal to the organism, as in plants which inherit genes inhibiting chlorophyll formation. Here genetics and biochemistry interact. The biophysicist and biochemist describe life as far as may be in physical and chemical terms, but many phenomena still elude this method. As Sherrington insists, bodily organs develop before their function can be used; the complex structure of the eye is built up before the eye can see.

Fertilization shows two steps—stimulation of the ovum and the union of the opposite nuclei, and stimulation can sometimes be performed parthenogenetically. If an ovum divides by falling into halves, it forms 'identical twins', while, if two ova are fertilized simultaneously, 'fraternal twins' result, and they may be no more alike than any two children of the same parents.

Heredity has also been examined further by the statistical study of large numbers initiated by Quetelet (p. 123) and complications due to mixture of different groups detected. Karl Pearson and others have traced the inheritance of qualities by measuring them in parents and offspring. If a group of men exceed the normal stature by four inches, their sons will on the average of large numbers exceed it by two inches—half as much. This relation is expressed by saying that the coefficient of correlation is one half or 0·5. If the sons had reverted to the normal, there would have been no correlation and the coefficient would be zero. Most hereditary qualities have coefficients of correlation ranging round 0·5. De Vilmorin, a member of a long-established family of French seedsmen, got better results in breeding plants by selecting as parents not from the qualities of individuals, but from those of lines which show continued good characters. The same holds true in breeding animals. It is a fallacy for a prospective bridegroom to say 'I am not marrying her family'. The children may be all too likely to show that he has perforce done so.

Though at one time there was controversy between Mendelians and Biometricians, in any complete study of heredity, as R. A. Fisher has shown, there is room statistically for both methods of inquiry.

The theory of evolution has become more and more firmly established as fresh geological evidence has accumulated. Some biologists still hold to Darwin's natural selection, acting on small variations, and some look to Mendelian mutations. But others feel that neither explanation gives an adequate cause for the present transmutation of species. It may be that living beings were more plastic in former ages, and are too fixed nowadays to show the fundamental changes necessary to produce new species, though the superficial modifications which give new varieties are still possible.

A survival of the fittest is no use to a nation unless the fittest have a preponderating number of children. An examination of books of reference in 1909 by my wife and myself showed that in England the landed, professional and upper commercial classes had diminished their output of children to less than one half of what it was in the years before 1870. An almost equal fall was shown by the statistics of Friendly Societies, whose members are mostly skilled artisans. On the other hand, miners and unskilled labourers were maintaining the numbers of their children, as were unfortunately the feeble-minded. It seems likely then that the differential birth rate is tending to breed out ability. Scholarships may supply the deficiency for a time, but the amount of ability in the country is limited. If it all be picked out and raised from the ranks, it may partially be sterilized by a decreased birth rate. Again, legislation, passed with other objects in view, may favour or depress special strains in the population. Death dues, for instance, are destroying the old landed families, on whom the nation has relied for unpaid work in the counties, and underpaid work in the Church, the Army and the Navy.

Much new evidence is now available about the effects of light and heat on the germination of seeds and the growth of plants. Maxima and minima of light and temperature are needed at some stages, and a process of 'vernalization' has been developed, whereby growing seeds are cooled for a time and the season of fruiting controlled. The subject of ecology, dealing with the relations of plants and animals with their inorganic environment and with other living creatures, is rapidly expanding.

Geology and Oceanography. Information on human evolution has recently been obtained by the discovery of fossil man-like apes and ape-like men. Early forms of anthropoid apes were found in the Egyptian Fayum and the foot-hills of the Himalayas, and in 1912 Dawson and Woodward discovered at Piltdown in Sussex man-like remains in Pleistocene deposits. It is possible too that the Chinese specimens named *Pithecanthropus Pekinensis* may be members of the group from which later types of men descend, while Neanderthal men were an aberrant line. Neolithic men came later, bringing with them into Western Europe traces of the world civilizations of Egypt and Mesopotamia. Considering fossils in general, Cambrian rocks, such as are found in North Wales, contain examples of most groups, but below Cambrian levels fossils fail. Somewhere between 500 million and 2000 million years ago, life appeared on the Earth—how we do not know.

The recent application of physical methods, such as an accurate measurement of gravity, has given information about regions of the Earth lying below the surfaces of the ground and sea. Seismic observations show that waves due to near earthquakes have travelled mainly through the crust of the Earth, while those from distant disturbances have traversed deeper regions. Indications suggest that the crust is shallow, perhaps 25 miles thick, while the central core, which may be chiefly liquid iron, has a radius more than half that of the Earth itself.

Artificial waves can be made by firing explosives, and, by noting the time of their echoes, the depth of the sea, or of discontinuities in the strata, or the distance of a buoy, can be found.

In the migration of fish there is usually a movement to a definite area for spawning, then a reverse movement in search of food. For instance, the salmon deposits its eggs in the upper reaches of rivers, the young move down to the sea, and when mature go back to the same rivers, thus showing individual memory. The European eel spends its adult life in fresh water, and migrates thousands of miles to spawn in the deeps of the Sargasso Sea. Many sea fish feed on diatoms and other minute organisms, collectively known as plankton; the drift of which shows where food and therefore fish will later be found.

Biochemistry, Physiology and Psychology. In 1912 (Sir) Frederick Gowland Hopkins showed that young rats, fed on chemically pure food, ceased to grow till minute quantities of fresh milk were added. Milk therefore contains substances necessary for growth and health, substances which Hopkins called 'accessory food factors', later known as vitamins. The most striking character of vitamins is the fact that minute quantities are enough to produce their beneficial effects. Those first investigated were given the letters A, B, C and D.

Vitamins A and D are present in animal fats, such as butter and cod-liver oil, and in green plants. A protects from infection; D is necessary for the calcification of bones and protection from rickets, and can be made by the action of ultra-violet light; B is found in the husks of various grains, in yeast, etc., and protects from diseases such as beri-beri and polyneuritis; C is present in fresh green plants and in fruits such as the lemon, and protects from scurvy.

Much information about vitamins has been found more recently. Vitamin A, proved by von Euler in 1929 to be allied to the vegetable pigment carotene, an unsaturated alcohol, is necessary for the health of the central nervous system, the retina and the skin; night blindness is an early symptom of vitamin A deficiency.

The first vitamin to be identified chemically was the anti-rachitic D; it is a complex alcohol known as ergosterol, and was isolated in 1927. Vitamin E maintains mammalian fertility; K is required for the normal coagulation of blood and so as a guard against the 'bleeding disease', haemophilia; chemically both these two are quinone derivatives.

Vitamin B has proved to be a mixture; the anti-neuritic B_1 or aneurin, found in yeast, etc., has been isolated in a crystalline form and proved to be a pyrimidine-thiazole compound. Some patients need the mass action of the pure B_1 for a cure from neuritis. Five other constituents of the B complex have been described. Vitamin C is ascorbic acid, a reducing compound $C_6H_8O_6$, allied to sugar. Its formation precedes chlorophyll and carotinoids in germinating seeds. In the animal body it is present in large amounts in two of the endocrine glands, the pituitary and the adrenal cortex. This is only a very short account of the many investigations which have been made since the discovery of vitamins.

Since Fischer synthesized mono-saccharide sugars (p. 106) investigations on the structure of di-saccharides such as cane sugar have been made by others, and ring formulae have been proposed. Fischer's work on amino-acids has also been followed up. But the most complex polypeptides yet prepared, with molecular weights somewhat exceeding 1300, are still far from proteins. By measurements of osmotic pressure and rates of sedimentation, proteins have been found to fall into two groups, with molecular weights that are simple multiples of 35,000 and 400,000 respectively. Indications of their structure have been given by X-ray examination, but no protein has yet been synthesized. This gap between the laboratory and life remains open.

X-ray photographs have also helped to explain in molecular terms the fibrous nature of such things as cellulose and the myosin of muscle (p. 135). Langmuir has related the constitutional formulae of organic compounds to their physical properties, and N. K. Adam has extended these methods to surface films. F. G. Donnan formulated in 1911 a theory of equilibria in membranes permeable to one kind of ion, which explains the phenomena of osmotic pressure and electric potential in such substances as the colloids of proteins.

The chemistry of blood includes the study of haematin, the structure of which has four rings linked by an iron atom, and combined with the protein globin to form the oxygen-carrying substance haemoglobin. In plants, Willstätter has shown that the nucleus of the chlorophyll molecule is similar to haematin with a magnesium atom replacing iron. He found two chlorophylls with slightly different composition, and in 1934 he gave diagrams of their structural formulae.

Oxidation in the tissues has been shown by Wieland to be carried on by specific enzymes, called dehydrogenases. They are present in all living tissues and able to liberate hydrogen to combine with oxygen. Muscular contraction was shown by Hopkins and W. M. Fletcher in 1907 to depend on the breakdown of glycogen to lactic acid. This process has since been analysed into eight chemical stages catalysed by at least ten enzymes.

Besides oxidation in the tissues there are also changes involving the addition of water and the shedding of amino-groups. The excretion of these bodies in the form of urea has been shown by Krebs to require a complicated cycle of reactions. Carbon dioxide is carried in the blood as bicarbonate, and Meldrum and Roughton have shown that it is released in the lungs by an enzyme, carbonic anhydrase.

In 1902 Bayliss and Starling found that pancreatic secretion is induced by a chemical substance formed in the intestine and carried to the pancreas by the blood. This led to the detection of other similar internal secretions, which Hardy named hormones (ὁρμάω, I rouse to activity). They are each secreted in one gland or organ and carried by the blood to others, for instance insulin, discovered in 1922 by Banting and Best by experiments on dogs, and used to obviate the effects of diabetes. The thyroid gland has already been considered (p. 116), but the work on other endocrines, that is hormones formed by glands of internal secretion, is so voluminous that it is now treated as a separate subject and called endocrinology. In this the co-ordinating rôle of the pituitary gland is specially important, particularly in the phenomena of sex.

The hormone adrenalin is discharged into the blood in conditions such as fright or anaesthesia, which stimulate the splanchnic nerves. Conversely, the injection of adrenalin produces the physical symptoms of emotion or fear.

In the study of the nervous system the pioneer work was done by Sir Charles Sherrington from 1906 onwards. Messages pass by the peripheral nerve fibres from the sense organs or receptors to the central nervous system and from it to the muscles and glands. But how are the incoming messages co-ordinated and the outgoing controlled in such a way that the animal responds as a whole with appropriate movements? Sherrington has shown that much of this 'integrative action' of the nervous system can be made intelligible by a study of the simple reflexes and their interaction. An incoming message may have a dual effect, exciting certain nerves and depressing or inhibiting others, sometimes helped by the adjustment of the time relations.

Experiments on nerve-impulses are facilitated by modern physical instruments. Single nerve-fibres can be examined, and minute heat effects have been detected by A. V. Hill, Gasser and Erlanger.

The highest part of the nervous system, the brain, is connected with sight and hearing, putting the animal into touch with distant objects. Mental functions have their seat in a part of the brain called the cerebrum, and especially in its cortex, which has been mapped out and its local reactions studied especially by Horsley, Head and Sherrington. Another part of the brain, the cerebellum, is connected with balance, posture and movement, acting in response to stimuli received from the muscles of the body and from the labyrinth of the ear.

The involuntary nervous system, which controls the unconscious bodily functions, was first studied thoroughly by Gaskell and Langley. Pavlov held that here psychological ideas were unnecessary. The simple unconditioned reflexes pass into complex reflexes conditioned by other factors, but the method of observing stimulation and resultant action may still be applied. This does not touch the problem of the nature of the intervening consciousness, but it has led, in the hands of Lloyd Morgan and J. B. Watson, to a school of psychology called behaviourism, which ignores consciousness in its investigations. Freud too held to a strict determinism, explaining our most trivial mistakes and our most cherished beliefs by the operation of instinctive forces which grow with the body and may cause mental ill-health if their development is checked or distorted.

All this work, culminating in behaviourism and Freud's determinism, has led to a mechanistic trend in recent psychology, which seeks to build up the individual from an accumulation of experiences and memories. But one may point out that to the behaviourist a man is only a nexus of stimuli and responses because the method of investigation by its own definitions and axioms is merely the study of the relations between stimuli and responses; it deliberately excludes consciousness and its problems. Some physiologists and philosophers point to purpose in the action of living beings as wholes, and the adaptation of embryonic organs to their final functions, as evidence for a modern form of vitalism. Perhaps we are still subject to the swings of the pendulum. The modern concept of organism recalls Aristotle's dictum that the animal body is not the mere sum of its parts.

Physiology and biochemistry are working their way into medicine. Besides the instances already mentioned, we may cite as other examples Minot's use of liver extract to cure pernicious anaemia, and the avoidance of miners' cramp by drinking salt water instead of fresh. Heavy labour carries away salt in the sweat, and fresh water dilutes the body fluids too much.

Viruses and Immunity. The first description of an ultra-microscopic, non-filterable virus was due to Ivanovski, who discovered one in tobacco-mosaic in 1892. Many others have since been found in both plants and animals. Small-pox, yellow fever, measles, influenza and the common cold, in cattle foot-and-mouth disease, in dogs distemper, in plants tulip-break, potato-leaf-roll and tobacco-mosaic are recognized as due to viruses.

The sizes of the virus particles have now been estimated in several ways. Filters can be made of collodion films with minute pores of regular size, measured by observing the rate of flow of water through a given area of film. Other methods depend on photography, on an ultra-violet or electron microscope, or on a high-power centrifuge. Particles range from 300 millimicrons, an approach to bacteria, to about 10 millimicrons in foot-and-mouth disease, a millimicron being the millionth of a millimetre.

Is the virus a minute living organism or a large chemical molecule? W. M. Stanley of Princeton obtained from tobacco virus a protein of high molecular weight which had the virus properties and also crystalline affinities, and other viruses are regular crystals. But they have some of the properties of living organisms; the diseases they cause are infectious, and the virus particles reproduce themselves in the new host. They seem to be borderline entities, and perhaps, as further evidence accumulates, they may throw light on the origin of life; but that is not yet.

The methods by which viruses travel are various. In an animal host, they may move through the blood, nerves or lymph, and the transmission from one host to another is often a complex process. Tobacco-necrosis is carried by an air-borne virus. Some viruses are carried by insects, such as the green-fly or thrips, while the viruses of louping-ill in sheep and red-water in cattle are conveyed by ticks. Kenneth Smith has found a plant disease to produce which two viruses are needed, one borne by insects and one otherwise.

An infectious disease is often found to make the patient immune from further attacks, the classical instance being the mild cow-pox which protects from the virulent disease (p. 116), probably by the formation of the same protective substances which are effective after small-pox itself. The body reacts to the injection of bacteria and many other proteins by making substances which can neutralize the poison; they are known as 'anti-bodies'. The substances causing this reaction are called 'antigens'. Heidelberger and Kendal have given some evidence of combination in definite chemical proportions between antigen and anti-body, but the action has also been explained as the union of oppositely charged colloidal particles.

Paul Ehrlich (1854–1915), who was responsible for much of the early work on immunity, produced in 1912 an arsenic compound, which he named 'salvarsan', with a specific destructive action on the micro-organism producing syphilis, and in 1924 Fourneau obtained a derivative of carbamide which destroys the parasite of sleeping sickness. In recent years a series of synthetic drugs, based on sulphanilamide, have been found to control the diseases in men and animals due to streptococci and pneumococci, and sulphaguanidine has been shown to be a specific remedy for dysentery.

Dunkin and Laidlaw found that the virus of distemper, weakened by formaldehyde, gave considerable immunity to dogs, confirmed by subsequent injections of virus. Some virus diseases, foot-and-mouth in cattle and influenza in man, show a variety of different strains, and immunity to one strain may not protect from others. This makes it difficult to secure protection.

Anthropology. Physical anthropology is, I suppose, a branch of natural history. Social anthropology touches on one side psychology and on others geography, sociology and comparative religion.

Mankind show marked differences in different parts of the world, the chief distinguishing feature being skin-colour, to which other characters are attached. The population of Europe has been divided into three races. In the north, and especially round the shores of the Baltic, are found tall, fair people with long-shaped skulls, who have been called Nordic. In the south, near the Mediterranean, and stretching up the lands of the Western Atlantic, are short, dark men also with long-shaped skulls, the Mediterranean race, while pushing between them are the Alpine people with Asiatic affinities, stocky, of medium colouring, with broad round skulls. The three races are only found in purity in small areas, but they are revealed by a general study, which shows a gradual approach to the type as we move towards its chief home.

It was formerly thought that similar civilizations might arise spontaneously in different parts of the world, but studies of various arts have favoured the alternative view that they often indicate a common origin. For instance, the widespread custom of erecting monoliths and other stone structures orientated in relation to the Sun and stars, which is seen from Egypt to Stonehenge, shows a connexion, not necessarily of race, but of civilization, perhaps due to the intermingling of peoples, or close intercourse in trade.

A vast collection of facts about primitive people in ancient and modern times has been put together by Sir James Frazer in his great book the *Golden Bough*, while the psychology of existing savage races has been studied by anthropologists like Rivers who have lived among them. Rivers introduced a new method; having found that the general terms in which former observers put their questions were quite unintelligible to the savage mind, he asked single questions and generalized afterwards. For instance, it is useless to ask a man if he can marry his deceased wife's sister. One must first inquire 'Can you marry that woman?' and then 'What relation are you to her and she to you?'

The old view of religion was that it was a body of doctrine, theology if the religion was one's own, or mythology if that of other people. Ritual was merely a form in which these beliefs were expressed publicly. Greek religion, for example, to the nineteenth century meant Greek mythology, though neither Greek nor Roman had any creed or dogma. But underlying the mythology were the Greek mysteries, and with them a system of ritual containing magical elements. Ritual is prior to and dominant over any definite belief. Compare this with some forms of modern psychology which say 'I react to outer stimuli and so I come to think.'

The relations of magic to religion and science are still uncertain. Some hold that magic is the common matrix out of which both religion and science arose, but Frazer thinks that they grew in series —magic, religion and then science. Rivers says that *mana*, a vague sense of awe and mystery, is a more primitive source of magic and religion than the animism described by Tylor. Magic is an attempt to get control over nature: the savage wants the Sun to shine, and so dances a Sun-dance. Sometimes he is related to some animal, which becomes endowed with sanctity—it may be *taboo* and must not be touched, it may be that by eating its flesh he will acquire its strength. A very widespread magic is that which, by rehearsing the drama of the year, hopes to secure fertility for crops, beasts and men.

When this fails, or otherwise ceases to satisfy, men turn to gods. At a high level, Tammuz of the Babylonians became Adonis of the Greeks. Tammuz is the spouse of Ishtar the great mother, goddess of fertility, and the union of Adonis and Aphrodite was necessary in Greece for the fertility of the Earth. These underlying rites of magic and mystery religions sought mystic union with nature or the divine through rites of initiation and communion. By eating the body of a god, corn for a corn god, the savage shares his attributes and powers. Drinking wine in the rites of the vine-god Dionysos is not revelry but a sacrament.

CHAPTER X. *THE NEW PHYSICS AND CHEMISTRY*

The Physical Revolution. During the last five years of the nineteenth century, a complete change in physical science began. The chemical atom, revealed by Dalton ninety years before, and accepted as the indivisible unit of matter, was shattered into fragments by J. J. Thomson and Rutherford. Then the facts led Planck to the theory that radiation is emitted in gushes or quanta, while Bohr and others imagined models of the atom in which Newtonian dynamics no longer held, and, models discarded, explanations had finally to be left in the equations of a new science of wave-mechanics. Newton's ideas of absolute space and time and his scheme of gravitational forces, which had replaced Aristotle's teaching and ruled mechanics and astronomy for two centuries, were superseded by Einstein; time and space became relative to the observer, and gravity a curvature in a space-time continuum.

The first shadow of the coming events was the accidental discovery by Röntgen in 1895 that covered photographic plates became fogged if electric discharges were passed through highly exhausted glass bulbs in their neighbourhood. Rays of some sort—let us call them X-rays—must be produced in the bulb, and pass through the glass and through the opaque coverings of the photographic plates. The tremendous benefits which this discovery has conferred on mankind through medicine and surgery will be known to all my readers.

When X-rays pass through a gas, they make it a conductor of electricity, and this process was jointly investigated by Thomson and Rutherford. If the rays were cut off, the conductivity persisted for a time and then died away. It was also removed at once by passing the gas through glass wool or between two plates oppositely electrified. The facts could be explained by supposing that the conductivity was due to ions as in liquids, but that in gases they had to be formed by some ionizing agency such as X-rays, and, if left alone, gradually recombined and neutralized each other.

Cathode Rays and Electrons. It has been known since 1869 that when a glass tube with platinum electrodes is very highly exhausted with an air pump, and an electric discharge passed through it, rays (since called cathode rays) fly off in straight lines from the negative electrode or cathode. It is these that give rise to X-rays where they strike solid objects. Cathode rays were thought to be either waves of the same nature as light or flights of negatively electrified particles, the latter alternative chiefly because they are deflected from their straight path by a magnetic force. Assuming the particle theory, the deflexion will be proportional to the electric charge and inversely proportional to the mass of the particles, that is proportional to e/m, the ratio of charge to mass. But it must also depend on the velocity v with which the particles move, so, taking e/m as one quantity, we have two unknowns, e/m and v, and shall want two equations to determine them. Several physicists, among them Wiechert and Kaufmann, measured the magnetic deflexion, and guessing, or getting indirect evidence of other properties, obtained values for e/m and v, the latter about one-tenth that of light, and the value of e/m much greater than for the hydrogen ion in liquid electrolysis. Next J. J. Thomson (1856–1940, Professor at Cambridge, later Sir Joseph Thomson, O.M.), directing the rays into an insulated cylinder, measured the negative charge they delivered and the heat developed when they struck a thermo-couple. The latter quantity gave the kinetic energy of the rays, which involves their velocity and supplies the second equation needed to determine both e/m and v.

But, in October 1897, Thomson got clearer results by a better method. If the cathode-ray stream consists of charged particles, they should be deflected by an electric as well as by a magnetic force. Attempts to get this deflexion had hitherto failed or proved inconclusive; but Thomson succeeded with the glass apparatus shown in the diagram. C is the cathode, A the anode, pierced by a slit which lets through a beam of rays which is further cut down by a second slit at B. The narrow pencil thus obtained fell on a fluorescent screen or photographic plate at p, after passing between two insulated metal plates at D and E. These plates, connected with the opposite poles of a high-tension battery, set up an electric force between them, and an electro-magnet, surrounding the whole apparatus, could be used to produce a magnetic field also acting on the rays. The electric deflexion from p to p' gave a second value for e/m, so that both e/m and v were determined. The measurements again showed that e/m was much greater than the corresponding value for the hydrogen ion in liquids, later work giving a ratio of 1837. Thus it became clear that either the charge was greater or the mass less than for hydrogen. Thomson thought that the mass must be less, but sought further proof.

Much indirect evidence had accumulated suggesting that the charge on a gaseous ion was the same as on a monovalent ion in liquids, and this was now made definite. C. T. R. Wilson had shown

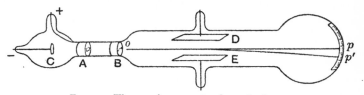

FIG. 14. Thomson's apparatus for cathode rays

that ions, like dust particles, act as cloud nuclei for the condensation of drops of water in moist air, and Thomson in 1899 used this action to find the electric charge carried. Then Townsend measured the rates of diffusion of ions through gases, and from his results again calculated the charge. All measurements agreed in showing that gaseous ions carry the same charge as monovalent ions in liquids.

Finally Thomson measured both e/m and v for the same ions—those ejected when ultra-violet light falls on a zinc plate—and got the same result. The proof was complete. Moreover, while the velocity varies, e/m is the same whatever be the nature of the electrodes or of the residual gas. The cathode ray particles, which Thomson called at first by the Newtonian name of 'corpuscles', are the common basis of matter sought by men from the Greeks onwards.

This discovery linked up with another line of research. On Maxwell's theory, light is an electro-magnetic wave, and must be emitted by vibrating electric systems. Lorentz therefore framed a theory of matter, in which parts of atoms are electric units, the vibrations starting light. This was supported by a discovery of Zeeman, who in 1896 found that lines in the spectrum of sodium were broadened by a strong magnetic field. From the broadening, it was possible again to calculate e/m. The same figure was obtained. To his electric unit Lorentz gave Johnston Stoney's name of 'electron', and later on this name was implicitly accepted by Thomson.

Thomson's picture of the atom was a uniform sphere of positive electricity, in which electrons are either at rest or revolving in planetary Newtonian orbits. When an atom loses an electron, the electron becomes a negative ion, while the residue is an atom positively electrified or a positive ion. One electron would place itself at the centre of the sphere, two at the opposite ends of a diameter, three at the corners of a triangle and so on, but with seven or eight the system becomes unstable, and a second shell of electrons is formed. This was held to explain chemical valency, and the recurrence of properties in the Periodic Table at each eighth element. But Thomson's sphere of positive electricity had to give way to Rutherford's theory of a nucleus, and Thomson's essentially Newtonian outlook pass into newer, less comprehensible concepts.

Radio-activity. X-rays produce marked effects on phosphorescent substances, and soon after Röntgen's discovery a search began for bodies which conversely might emit any sort of rays. In 1896 Henri Becquerel found that salts of uranium, and uranium itself, continually give out rays which affect a photographic plate through black paper and other screens, and were soon afterwards found to act like X-rays in making gases conductors of electricity.

In 1900 M. and Mme Curie made a systematic search of chemical elements and compounds and of natural substances. They found that pitchblende and other uranium minerals were more active than the element. Using radio-activity itself as a test and guide, they separated by chemical means a surprisingly active substance which they called radium. Other observers followed with elements which were named polonium and actinium. The amount of radium in pitchblende is very small, many tons of the mineral yielding only a small fraction of a gram of a salt of radium.

In 1899 Rutherford (see Plate VII, facing p. 123) of Montreal, later of Manchester, and finally Lord Rutherford of Nelson, O.M., Sir J. J. Thomson's successor at Cambridge, discovered that the radiation from uranium consists of three kinds of rays to which he gave the Greek letters α, β and γ. The α rays, the most powerful at short range, will only go through thin screens; the β rays through half a millimetre of aluminium before their intensity is halved, and the γ rays through much more. By measuring their magnetic deflexion α rays have been found to be atoms of helium carrying a double positive charge, β rays negative electrons, while γ rays have proved to be waves like light, but with enormously shorter wavelengths.

Crookes found that uranium, precipitated from solution by ammonium carbonate, gave a small quantity of an active substance which he named uranium X, the residual uranium being for a time less active. Rutherford and his colleagues discovered many such chemical changes, yielding from radium and thorium, among other products, radio-active gases which they called emanations. Radio-activity was found always to be accompanied by chemical change, new elements appearing and α rays, or β and γ rays, being shot off. The pedigree of the family beginning with uranium and containing radium, has been traced for fifteen generations before it ends in the inactive element lead.

The rate of decay in activity of these bodies varies enormously. Uranium would take $4 \cdot 5 \times 10^9$ years to fall to half value, radium 1600 years, radium emanation $3 \cdot 82$ days, radium A $3 \cdot 05$ minutes, and radium C 10^{-6} second. Some give out only α rays and others β and γ rays. Each single change proceeds so that the rate of decay during each short interval of time is proportional to the activity at the beginning of the interval. This is the same law of change as that shown by a chemical compound dissociating molecule by molecule into simpler products.

In 1903 Curie and Laborde discovered the remarkable fact that the compounds of radium continually emit heat, and calculated that one gram of radium would give about 100 gram-calories of heat per hour. They found the emission of heat was not changed by high or

THE NEW PHYSICS AND CHEMISTRY

low temperatures. The amount of energy thus liberated is many thousand times more than that of the most violent chemical action known.

Thus we have chemical changes of monomolecular type, giving a new element at each change and an enormous output of energy. Taking all these properties into account, Rutherford and Soddy in 1903 explained the facts by the theory that radio-activity is due to an explosive disintegration of the elementary atoms, in which an atom here and there out of millions explodes, flinging off an α particle, or a β particle and a γ ray, leaving a different atom behind.

The atomic theory from Democritus to Dalton had no hope of dealing with atoms singly; it could only treat them statistically in large, indeed gigantic, numbers; the kinetic theory of gases indicates that the number of molecules in one cubic centimetre is about $2 \cdot 7 \times 10^{19}$. But radio-activity gave several ways of detecting the presence of single atoms. Firstly, Crookes observed with a magnifying lens scintillations on a fluorescent screen of zinc sulphide exposed to a speck of radium bromide shooting out α rays, that is atoms of helium; each atom as it strikes the screen gives a flash of light. Secondly, Rutherford counted the kicks of an electrometer needle as α particles shoot through a gas, producing ions as they pass. Thirdly, C. T. R. Wilson used α rays to ionize a moist gas, the ions forming nuclei of condensation, so that each α particle gives a cloud-track in the gas. From the number of α particles the 'life' of the radium, that is the time it takes to halve its activity, can be calculated.

The cloud-tracks of α rays are usually straight, but sometimes a sharp change in direction is seen (see Plate VIII, facing p. 148). The forces exerted by electrons on the much more massive α particle cannot be enough to produce such a turn, but it is explained at once if we imagine with Rutherford that the positive electricity, which Thomson supposed distributed evenly over a comparatively large sphere, is concentrated in a very dense minute nucleus at the centre. If the atom is that of hydrogen, the nucleus, which is then called a proton, has about 1837 times the mass of an electron. This heavy nucleus, with the mass of an atom, is massive enough to stop and turn a colliding α particle of the same order of mass. The hydrogen atom has one electron outside the nucleus. At first, with Newtonian preconceptions, it was thought that the electron circled round in a planetary orbit; this idea has been abandoned, but before describing its successor we must follow other lines of research.

X-Rays and Atomic Numbers. The X-rays discovered by Röntgen are not refracted like ordinary light, and show little trace of reflexion or polarization. Unlike cathode rays or α and β rays, they are not deflected by magnetic or electric forces. But in 1912 Laue suggested that if X-rays were light of very short wave-length, they might be diffracted by the regular layers of atoms in a crystal, as light is diffracted by the regular scratches on a grating (p. 101), and a spectrum might result. This suggestion was verified experimentally by Friedrich and Kipping, and carried farther by Sir William and Sir Lawrence Bragg. Beginning with rock salt, a simple cubic type of crystal, the Braggs found spectral lines on a more diffuse background, lines which showed that the distance between the layers of atoms was $2 \cdot 81 \times 10^{-8}$ centimetre, and that the characteristic X-rays emitted from a target of palladium had a wave-length of $0 \cdot 576 \times 10^{-8}$ centimetre, only the one ten-thousandth part of the wave-length of sodium light. Electro-magnetic radiation is now known from the longest waves used in wireless telegraphy to the short waves of X and γ rays, a range of about 60 octaves, of which only about one octave is visible light.

The work on sodium and potassium chlorides showed no trace of NaCl or KCl molecules; the crystals are made of alternate positive and negative ions. Extension to more complex inorganic bodies confirmed this view, the silicates for instance are silicon-oxygen skeletons enclosing positive ions. But organic chemistry is properly based on the concept of the molecule, and X-ray analysis has shown that the structural formulae of the chemist are correct. Again, the crystals of metals and alloys have been analysed, and several observers (Herzog, Astbury, Bernal, etc.) have examined complex biochemical substances such as cellulose, keratin and proteins.

When the target bombarded by cathode rays was changed from

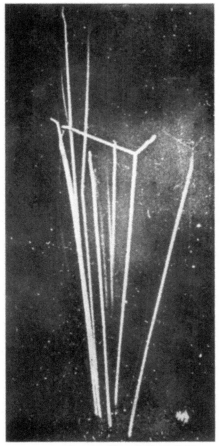

Photo. by P. M. S. Blackett

Plate VIII. Tracks of α particles in oxygen, one showing a fork due to collision with an oxygen nucleus. The short branch of the fork was produced by the recoiling oxygen nucleus and the long branch by the deflected α particle. Measurements of the angles of deflection of the two branches showed that momentum and energy were conserved in the collision.

Plate IX The Spiral Nebula in Canes Venatici

THE NEW PHYSICS AND CHEMISTRY

one metal to another, and the X-rays examined by using a crystal of potassium ferro-cyanide as a grating, H. G. J. Moseley found that the square root of the frequency of vibration (n) of the strongest characteristic line in the spectrum increases by the same amount on passing from element to element in the Periodic Table. If $\sqrt{n}$, that is $n^{\frac{1}{2}}$, be multiplied by a constant adjusted to bring this step-by-step increase to unity, Moseley got a series of atomic numbers for solid elements from aluminium 13 to gold 79. Fitting in the other known elements, it was found that, from hydrogen 1 to uranium 92 (atomic weight 238·2), there were only four or five gaps for undiscovered elements, and some of these have since been filled. It is sad to record that the brilliant young physicist Moseley, soon after this great discovery, was killed in the first world war—an incalculable loss to England and to science.

Positive Rays and Isotopes. Cathode-ray particles are carriers of negative electricity, or rather perhaps are themselves disembodied negative electrons. In an exhausted tube the anode also is a source of rays. If holes are bored in a cathode placed opposite the anode, the rays go through the holes and can be detected on the far side. Their magnetic and electric deflexions were measured in 1898, first by Wien and then by Thomson, and showed that they were particles carrying positive charges and had masses comparable with those of ordinary atoms and molecules.

In 1910–11 Thomson used a large vessel very highly exhausted with a long narrow tube through the cathode. This gave a very small pencil of rays which was recorded on a photographic plate inside the apparatus. The magnetic and electric forces were arranged at right angles to each other. The magnetic deflexion is inversely proportional to the velocity of the particles and the electric deflexion inversely to the square of the velocity. Hence, if identical particles of differing velocities exist in the rays, a parabola will be photographed on the plate (p. 59). The lines which appear depend on the nature of the residual gas in the apparatus. With hydrogen the fundamental line gives a value of 10^4 for e/m, or 10^{-4} for m/e, the same as for the hydrogen ion in liquid electrolytes. A second line has double this value for m/e, and indicates a hydrogen molecule with twice the mass of the atom carrying a single electric charge.

With the gas neon, to which chemical measurements give an atomic weight of 20·2, Thomson found two lines, indicating weights of 20 and 22. This suggested that neon as usually prepared was a mixture of two elements, identical in chemical properties but of different atomic weights. Such elements also appear in radio-activity, and were called by Soddy 'isotopes' (ἴσο-τόπος, occupying the same place in the chemical table).

Thomson's experiments were carried farther by F. W. Aston, who obtained regular 'mass spectra' of many elements. Taking the atomic weight of oxygen as 16, the atomic weights of all other elements are very nearly whole numbers; chlorine, for instance, which chemically is 35·46, was shown to consist of a mixture of two isotopes 35 and 37. When Aston's first apparatus (now in the Science Museum at South Kensington) was brought into action, results poured out. In 1933 Aston said: 'At the present time out of all the elements known to exist in reasonable quantities, only eighteen remain without analysis', and by 1935 about 250 stable isotopes were known. For practically every atomic weight up to 210 an element has been found; thus Prout's hypothesis has been confirmed. Some numbers harbour more than one element—'isobars', atoms of the same weight but different chemical properties.

To the hydrogen nucleus or proton, Aston gave a mass of 1·0076 now corrected to 1·00837. In 1932 Urey, by a process of fractionization, discovered that a heavy isotope of hydrogen with mass 2, double the normal detected by Aston, was present in ordinary hydrogen to the amount of 1 in 4000. It is now called 'deuterium' and its ion a 'deuteron'. By electrolysing water, Washburn obtained a new substance, heavy water, in which hydrogen is replaced by the isotope. Heavy water was isolated by Lewis; it is about 11 per cent denser than ordinary water.

The Structure of the Atom. If electrons in the atom revolved in Newtonian planetary orbits, they would emit radiant energy, and the orbit would contract with a consequent quickening of the period of rotation and of the frequency of the emitted waves. Atoms in all stages would exist, and therefore in all spectra radiation of every frequency should be found, instead of only the radiation of a few definite frequencies as seen in the line spectra of many chemical elements.

To meet these difficulties, Planck in 1901 devised a quantum theory, according to which radiation is not continuous but is emitted in gushes, so that, like matter, it exists only in indivisible units or atoms, not all of the same size, but with sizes proportional to the frequency of the radiation. High frequency ultra-violet radiation can only radiate when it has a large amount of energy available, so that the chance of many units being radiated is small, and the total amount is small also. On the other hand, the low frequency infra-red radiations possess small units, and the chances favour their emission, but, as they are small, the total amount radiated is again small. For some special range of intermediate frequency, where the unit is of intermediate size, the chances may be favourable to a maximum emission of energy. Thus the quantum theory, like many physical and chemical problems, is an exercise in probability.

In 1923 Compton put forward the idea of a unit of radiation comparable with the electron and proton; he called it a photon. The quantum of energy E is proportional to the frequency ν, that is inversely as the period of vibration T. So

$$E = h\nu = \frac{h}{T},$$

where h is Planck's constant, $= ET$, that product of energy and time which is called 'action'. This is a true natural unit, independent of anything variable, like the atom or the electron.

The quantum theory was much strengthened when it was also applied successfully to explain the variation in specific heat. Einstein pointed out that the rate of absorption would depend on the size of the unit, and therefore on the frequency of vibration and so on the temperature. This application has been carried farther by Nernst and Lindemann[1] and by Debye.

· Now Lord Cherwell.

The chief difficulty in quantum theory is to explain the interference between parts of a beam of light which seems to require a long train of uniform waves. The most hopeful attempts to reconcile the difference are found in a combination of waves and particles, for which there is experimental evidence, and in a recent theory of wave-mechanics.

The application of quantum theory to the problem of atomic structure was first made by Niels Bohr of Copenhagen, when working in Rutherford's laboratory in Manchester. In the complex spectrum of hydrogen, regularities appear if we examine, not the usual wavelengths of its luminous lines, but the number of waves in a centimetre, and Bohr was able to explain these regularities on the quantum theory.

If action is absorbed in quantum units, only a certain number of all the conceivable orbits will be used by the electrons. In the smallest orbit the action will be one unit or h, in the next $2h$ and so on. If an electron leaves one path, it must jump instantaneously to another, apparently without passing through the intervening space. Here, for the first time, we leave Newtonian dynamics, and open a new chapter in physical science.

By assuming four possible orbits for the single hydrogen electron, Bohr explained the facts of its coarser spectrum. But Bohr's atom failed to account for the finer details of the spectrum of neutral helium and the great complexity of the spectra of heavier elements. By 1925 it was becoming clear that Bohr's theory, so successful for a time, was inadequate.

Nevertheless the idea of different energy levels is supported by the facts of ionization. Lenard in 1902 showed that an electron must possess a certain minimum energy before it would ionize a gas. Franck and Hertz found that maxima of ionization occur at multiples of a definite voltage. At the same points new lines or groups of lines may appear in the spectra.

The double aspect of electrons as particles and waves was demonstrated experimentally in 1923 by Davisson and Kunsman and in 1927 by Davisson and Germer working in America, and later in that year by Sir G. P. Thomson, son of Sir J. J. Thomson. He passed a narrow beam of electrons through an exceedingly thin sheet of metal, thinner than the finest gold leaf. A photographic plate on the far side showed a series of diffraction rings like those obtained when light is passed through a thin glass plate or a soap film. This indicates that a moving electron is accompanied by a train of waves, the wavelength being found to be about the millionth part of that of visible light, waves and electrons must vibrate in unison, hence, even experimentally, the electron has a structure and more minute parts. Mathematical investigation proves that the energy of the electron is proportional to the frequency of the oscillations, and that the product of the momentum of the electron and the wave-length is constant. It will be seen with astonishment how closely this latest concept of electron and attendant wave-train resembles Newton's theory of light, with its corpuscles and waves that put them into "fits of easy reflexion and easy transmission".

We can only examine atoms from outside, observing the radiation or radio-active particles which enter or leave; we cannot say that any given internal mechanism, such as that imagined by Bohr, is the only mechanism that will produce the external effects. But in 1925 Heisenberg framed a new mathematical theory of quantum mechanics, based only on the frequencies and amplitudes of the emitted radiation and on the energy levels of the atomic system. The theory gives the main lines of the hydrogen spectrum and the influence on it of electric and magnetic fields.

In 1926 Schrödinger extended certain work of de Broglie on high quanta. Taking material particles as wave-systems, he got equations mathematically similar to those of Heisenberg. The velocity of a single wave is not the same as the velocity of a group of waves or storm, which appears to us as a particle, while the frequencies manifest themselves as energies. Thus we return to the constant relation between frequency and energy first seen in Planck's constant h. Schrödinger obtained concordances between his equations and the spectrum lines even in complex atoms.

When one of the wave-groups is small, there is no doubt where to place the electron which is its manifestation. But, as the group expands, the electron can be put anywhere within it, so that there is an uncertainty in its position. The more accurately we attempt to specify the position of a particle, the less accurately can the velocity be determined and *vice versa*. The product of the two uncertainties, approximately at any rate, brings us back once more to the quantum constant h. Thus twenty-five years after the atom was resolved into electrons, electrons themselves were resolved into an unknown type of radiation or into a disembodied wave-system. The last trace of the old, hard, massy atom has disappeared, mechanical models of the atom have failed, and the ultimate concepts of physics have, it seems, to be left in the decent obscurity of mathematical equations.

The Transmutation of Elements. For some years after the acceptance of the atomic explosion theory of radio-activity, all attempts to initiate or control radio-active transformations failed. But in 1919 Rutherford discovered that bombardment with α rays induces atomic changes in nitrogen, with the emission of fast-moving hydrogen nuclei or protons, the existence of which was confirmed by Blackett, who photographed their paths in a cloud chamber. This discovery began an immense development in controlled atomic transformations. Between 1921 and 1924 Rutherford and Chadwick disintegrated most elements from boron to potassium.

When beryllium was bombarded, Bothe obtained rays even more penetrating than the γ rays of radium, and in 1932 Chadwick found that the main part of this radiation consisted, not of γ waves but of swift uncharged particles about equal in mass to hydrogen atoms. They are now called neutrons, and, being uncharged, pass freely through atoms without causing ionization. But, as Feather, Harkins and Fermi have shown, neutrons, especially slow neutrons, entering easily into nuclei, are very effective in causing transmutation.

Yet another type of radiation, always passing through space, has been detected. The charge on an electroscope leaks away faster when it is raised above the Earth's surface and more slowly when sunk in water, the radiation being more penetrating than any terrestrial ray. The intensity is the same day and night, and the rays arrive in the southern hemisphere when the Milky Way is not visible. Thus they cannot come from Sun or stars, but must originate in outer space. Their energies were measured by Carl Anderson and Millikan, who passed them through an intense magnetic field and measured the deflexion, finding energies ranging round 6 thousand million electron-volts, the electron-volt being the energy change when one electron falls through a potential difference of one volt. In 1932 Anderson discovered positive particles with the mass of negative electrons. These particles are now called positrons.

When photons strike the nucleus of a heavy atom, a positive-negative electron-pair appears in a cloud chamber. Their joint energy is about 1·6 million electron-volts when the energy of the photons was 2·6 millions. The difference of 1 million e-volts measures the energy involved in the conversion of photons of radiation into an electron-pair of matter. Conversely if a positron and an electron annihilate each other, two photons of electro-magnetic radiation shoot out in opposite directions. In 1938 Anderson and Neddermeyer confirmed a supposition that highly penetrating particles exist having masses intermediate between electrons and protons—about that of 200 electrons, a proton being almost 2000. The intermediate particles have been called mesotrons. The mode of origin of the cosmic rays is still uncertain.

Thus it will be seen how complex the structure of matter must be; we have the following types of particles:

Name	Mass in electron units	Electric charge
Electron or β particle	1	$-e$
Positron	1	$+e$
Mesotron	200	$\pm e$
Proton	1800	$+e$
Neutron	1800	0
Deuteron	3600	$+e$
α particle	7200	$+2e$

Besides these particles, which are reckoned as material, there is the photon, the unit of radiation. It is clear that matter is a wonderful and mysterious thing, too complex to be represented in atomic models, so that, as we have seen, its description may have ultimately to be left in a series of wave-equations.

The quantity of α rays obtained from radio-active substances is very small, and artificial methods of producing more effective α rays, long sought, have now been found. By passing an electric discharge through hydrogen or its isotope deuterium, a copious supply of protons and deuterons can be obtained, but, to give them velocity, a very strong electric field is necessary. This was applied by Cockcroft and Walton, and now powerful apparatus is available. Again, E. Lawrence of California has invented a new form of accelerating apparatus called a 'cyclotron'. Ions pass through an alternating electric field and a magnetic field at right angles, which makes the proton or deuteron describe a spiral path of steadily increasing radius, and enter and leave the electric field at regular intervals. For one frequency of the alternating potential, the ions always enter the field when the electric force is in the direction to accelerate them. In this way Lawrence got intense streams of protons and deuterons, equivalent to the α rays from 16 kilograms of radium.

With these instruments many transmutations have already been performed. To take an example, Lawrence and his colleagues, by bombarding bismuth, have converted it into a radio-active isotope identical with the natural radio-active element radium E, which has a time of half-decay of 5 days. Again, sodium (atomic weight 23) or its salts give a radio-active isotope of weight 24. This radio-sodium changes with the emission of one β particle, into magnesium, also 24, the period of half-decay being 15 hours. In the course of these recent researches more than 250 new radio-active substances have been recorded. In such transmutations as these, the dream of the mediaeval alchemist has come true.

Some of the energy changes in these forced transmutations are greater than those in natural radio-activity. For instance, a deuteron of energy 21,000 electron-volts will transmute an atom of lithium into beryllium with an emission of energy of 22·5 million electron-volts. This looks at first sight like a limitless source of atomic energy. But only about one deuteron in 10^8 is effective, so that, on balance, more energy has to be supplied than is emitted. A similar catch appears in every case yet examined, and, even if it could be obtained, it might be dangerous to put the destructive power of atomic energy

THE NEW PHYSICS AND CHEMISTRY

into the hands of man. Nevertheless, the advance of knowledge cannot be stopped, and science is not responsible if a bad use is made of some of the gifts it brings.

Electro-magnetic Waves. Clerk Maxwell published his electro-magnetic equations in 1864, and Hertz produced and demonstrated electro-magnetic waves in 1887 (p. 103). Rutherford and others carried the work farther in later years, sending signals over a mile or two, but it only led to a practical outcome when two inventions were adopted. Marconi used an aerial wire or antenna to increase the energy despatched and collect that energy at the receiving station, and Sir O. W. Richardson investigated the emission of electrons from hot metals and so made possible the thermionic valve.

Hertz and other early experimenters used the waves from an induction coil, waves which are heavily damped and rapidly die away, but for radio-transmission a train of continuous, undamped waves is needed. If a hot wire inside an exhausted bulb be connected with the negative terminal of a battery, and a metal plate with the positive terminal, a continuous negative current, carried by the electrons, will pass from wire to plate, though if the terminals be reversed no appreciable current will flow. Thus the thermionic valve acts as a rectifier, letting one half of an alternating current pass and stopping the other half. If a grid of wire gauze be put between wire and plate and be positively electrified, it will help the emission of electrons and increase the thermionic current, but, if it be negative, it will decrease it. If it alternates in potential, the current also will oscillate, an alternating current being thus superposed on a direct one. These alternations are passed through the primary circuit of a transformer, the secondary being connected back to give the grid its proper alternating potential. The thermionic valve is used both to emit a steady, undamped train of waves and also to rectify them when received. By interrupting these rectified currents and passing them through a telephone, a sound of corresponding pitch is obtained and radio-speech made possible.

The radiation can be divided into an earth wave, gliding over the surface of the ground, and a sky wave, starting above the horizontal.

The latter wave can be carried long distances, because it is reflected or refracted by a conducting layer in the upper atmosphere, ionized by the Sun's rays. This is called the ionosphere, or the Kennelly-Heaviside layer from those who first suggested its existence. From the behaviour of long-distance waves, much information about the ionosphere has been obtained. It was first located by wireless waves by Appleton and Barnett, and in 1926 an upper layer was discovered by Appleton, and shown to be the more effective in deflecting short radio-waves. These experiments were an early instance of radio-location.

Another invention using modern discoveries is the electron microscope. The ordinary microscope fails to give clear definition when the size of the object approaches the wave-length of light. The ultra-microscope (see p. 111) carries vision somewhat farther, but the train of waves accompanying an electron has wave-lengths about the millionth part of that of light.

The function of the lenses in a microscope is to bend the rays of light coming from the object and focus them for the eye. Highly accelerated electrons can similarly be bent by a properly adjusted magnetic field and focused on a fluorescent screen or photographic plate inside the apparatus. Objects much smaller than those otherwise visible are thus clearly seen. The minute beings which devour bacteria have been revealed, and, in the near future, a great extension of knowledge will thus be won. Viruses should easily be seen, and a vision of molecules or even atoms seems within the bounds of possibility.

Relativity. In 1676 Römer discovered that light needed time for its propagation by observing the eclipses of the satellites of Jupiter when the Earth was approaching or receding, and found a velocity of 192,000 miles a second. In 1728 Bradley again measured the velocity by observing the aberration of light from the stars as the Earth moves in its orbit. In 1849 Fizeau passed a beam of light through one of the blanks in a toothed wheel and reflected it back from a mirror 3 or 4 miles away. When the wheel was at rest, the return beam passed through the same gap, but, when the wheel was revolving fast enough, the beam struck the next tooth and became invisible. Next Foucault measured the velocity by means of a rapidly rotating mirror. The best modern results give a value of 186,300 miles, or $2 \cdot 998 \times 10^{10}$ centimetres a second *in vacuo*, very nearly 3×10^{10}.

If there be a luminiferous aether, its effect on light should disclose its motion. If the Earth moves through the aether without disturbing it, Earth and aether will be in relative motion, and light should travel faster with the aether stream than against it, and faster to and fro across the stream than up and down it. On these lines Michelson and Morley in 1887 designed an apparatus to test the problem. They could find no difference when the light took one of the alternative paths; in this experiment there was no appreciable relative motion of Earth and aether; the Earth seems to drag the aether with it.

But, in calculating the velocity of light from aberration, it is assumed that the aether is undisturbed by the passage of the Earth. Moreover in 1893 Lodge rotated two parallel heavy steel plates and passed light between them. He could find no change in the velocity of the light when the plates were at rest and when they were rotating rapidly. Thus there is an unresolved discrepancy.

G. F. Fitzgerald suggested that if matter be electrical in essence, or even bound together by electric forces, it might contract as it passed through an electro-magnetic aether. The necessary contraction is very small, and could not be observed with scales, because they too would contract. Michelson and Morley's apparatus might change in dimensions so as to mask the effect expected. Whatever the cause, every attempt to find a change in velocity failed.

But in 1905 an entirely new direction was given to thought on this subject by Albert Einstein, who pointed out that the ideas of absolute space and time were metaphysical concepts and did not necessarily follow from the observations and experiments of physics. The Fitzgerald contraction would be invisible to us moving with the scales and suffering corresponding changes, but it might be measurable by an observer moving differently. Time and space then are not absolute but relative to the observer, and are such that light always travels relatively to any observer with the same velocity. That is the first experimental law of the new physics.

The mass of a moving body will increase in the same proportion as its length in the direction of motion is shortened. It can be shown that, on the principle of relativity, $m = m_0 \big/ \sqrt{1 - \frac{v^2}{c^2}}$, where v is the velocity of the body, c that of light, and m_0 the mass when at rest. This result can be tested by observations on β particles and has been shown to agree with the measurements by Kaufmann. Again, on the principle of relativity, mass and energy are equivalent, a mass m expressed as energy being mc^2. This too is in conformity with Maxwell's theory of waves, which possess momentum equal to E/c, where E is their energy. Momentum being mc, we get again $E = mc^2$.

Space and time then are relative to the observer; but in 1908 Minkowski pointed out that space and time compensate each other, so that a combination of the two, where time is a fourth dimension to the three of space, is the same for all observers. Other quantities that remain absolute are number, thermodynamic entropy and action, that product of energy and time that gives us the quantum. In reversible physics events can occur in either direction, but in the second law of thermodynamics, and the irreversible rise of the entropy of an isolated system towards a maximum, we have a physical process which can only proceed in one direction, like the remorseless march of time in the human mind.

In Minkowski's space-time there are natural paths like the straight paths of freely moving bodies in three-dimensional space. Since a projectile falls to the Earth and the planets circle round the Sun, we see that near matter these natural paths must be curved, and there must be a curvature in space-time.

Calculation shows that the consequences of this theory are the same as those of Newton's to the usual order of accuracy of observation. But in one or two cases it is just possible to devise a crucial experiment. The deflexion of a ray of light by the Sun is twice as great on Einstein's theory as on Newton's. During the eclipse of 1919, Eddington in the Gulf of Guinea and Crommelin in Brazil photographed the image of a star just outside the Sun's disc. Compared with stars farther away from the Sun the image of the near star was displaced to the amount foretold by Einstein. Secondly a discrepancy of 42 seconds of arc per century in the orbit of Mercury left by the Newtonian theory was at once explained by Einstein, who calculated a change of 43 seconds of arc. Eddington has linked gravitation with electricity and quantum theory by comparing the theoretical with the observed values of many physical constants. He obtained very striking concordances. It seems that all these modern concepts may before long be brought together in one new physical synthesis.

CHAPTER XI. *THE STELLAR UNIVERSE*

The Solar System. Newtonian astronomy was chiefly concerned with the solar system, minute compared with the stellar universe, but of special interest to ourselves. Kepler gave us a model of the system, but the scale of the model was not known till some one distance was measured in terrestrial units. For instance, when the planet Venus passes between the Earth and the Sun, its time of transit over the Sun as observed at two places on the Earth gives a means of determining by trigonometry the distance of the Sun. This proves to be 92·8 million miles, which light can traverse in 8·3 minutes. The most distant known planet, Pluto, discovered by Tombaugh in 1930, moves round the Sun in 248 years, the diameter of the orbit being 7350 million miles, which gives the size of the solar system as now known.

There has often been discussion on the possibility of life on other worlds; this problem reduces to a consideration of the conditions on the other planets of the solar system. One of the most important of these conditions is the nature of the atmosphere round each planet. Molecules of the gases will escape from gravitational attraction if their velocity of movement is more than a critical value called the 'velocity of escape', which can be calculated from the mass and size of the planet. The moon has lost practically all its atmosphere, Mars and Venus have atmospheres comparable with that of the Earth, while the large planets—Jupiter, Saturn, Uranus and Neptune— have much more than the Earth. On Venus carbon dioxide is plentiful; but apparently there is no vegetation and no oxygen, and life is not yet possible, whereas on Mars conditions favourable to life are over, or, at all events, drawing to a close. Planets much farther from the Sun than the Earth are probably too cold to support life.

The Stars and Nebulae. Beyond the orbit of the planet Pluto lies a great gulf of space. The few nearest stars may be seen to move against the background of those more distant as the Earth passes in six months from one side of its orbit to the other. In 1832 and fol-

THE STELLAR UNIVERSE

lowing years, measurements of this parallax were made. The nearest star, a faint speck called *Proxima Centauri*, was found to be 24 million million ($2 \cdot 4 \times 10^{13}$) miles away from us, a distance traversed by light in 4·1 years, and 3000 times the diameter of Pluto's orbit.

The naked eye can see a few thousand stars. They were classed by Hipparchus in six 'magnitudes' according to their apparent brightness. For those whose distances are known, we can calculate the apparent magnitude the star would have at a standard distance and this is called the absolute magnitude. In the 100-inch reflecting telescope at the Mount Wilson Observatory in America, a number of stars estimated at about 100 million are visible, and hundreds of times more must exist. There are many double stars, first discovered by Herschel; some are too near each other to be separated by a telescope, but can be resolved spectroscopically by observing the shift in the spectral lines as the two stars are alternately approaching or receding from us (p. 101). If both stars of the doublet are luminous, the lines are doubled. If one is invisible, sometimes it periodically hides the other, and thus again the double nature of the pair can be detected. If visual and also spectroscopic measurements are possible, a very complete specification can be obtained, giving the individual orbits and masses. Other variable stars cannot be explained by eclipses, as, for instance, δ Cephei. These 'Cepheid' stars show a relation between the period of variation and the luminosity or absolute magnitude. This relation is so regular that measurement of the periods of other similar stars at unknown distances can be used to estimate their absolute magnitudes. An observation of the apparent magnitude then gives the distance—a method applicable to stars too far away to show any parallax.

Stars are most numerous in a band stretching across the heavens in a great circle. It is called the Galaxy or Milky Way. We and our Sun are within it but not at the centre. This, our stellar system, forms a vast lens-shaped collection of stars of which the Sun is one. The size is so huge that light would take 300,000 years to traverse the longest diameter. In the galaxy there are two great streams of stars moving in different directions, while the galaxy itself is rotating about a centre. The stars of which the masses can be determined do not seem to differ much, mostly ranging from one half to three times the mass of the Sun.

Beyond our stellar system lie other galaxies, 'island Universes', of which our own is but one. Most impressive of all are the multitudes of great spiral nebulae, star-systems in the making. They are gigantic; though formed of tenuous gas, one of them might form a thousand million Suns. Estimates of their distances give to the farthest 500 million light-years; they lie far beyond the confines of our stellar system (see Plate IX, facing p. 149).

The Structure of Stars. A system of classification of stars by their spectra was introduced by Father Secchi in Rome about 1867, and has been much extended and improved at the Harvard Observatory in America. A list has been drawn up in which the bluer stars come first, and the different types of spectra, indicating various elements and temperatures, are grouped together. Again, if a black body, taken as a perfect radiator, is heated, the character as well as the intensity of the radiation changes. For each temperature there is a characteristic curve between radiant energy and wave-length, showing a maximum at one particular wave-length. As the temperature rises, the position of the maximum shifts towards the blue, and thus the temperature can be estimated. The temperatures of the outermost radiating layers of stars range from 1650° C. to 23,000° C.

Knowing the size and average density of the Sun or a star, and assuming that it is gaseous, Eddington calculated the rate of increase of pressure with depth below the surface. At any level, the pressure from above is supported by the elasticity of the gas below and the pressure of its radiation. These quantities are known and depend on the internal temperature, and so the temperature can be estimated. To support the enormous pressure within the Sun or another similar star, the internal temperature must be very high—reaching tens of millions of degrees Centigrade. Sir R. H. Fowler has shown that in stellar atoms some electrons must stay in orbits of high energy; the pressure required to decrease the volume is then increased, and the internal temperature needed to balance the pressure is less; at the centre of the Sun about 20 million degrees. These figures are, of course, altogether different from the temperatures of the outermost radiating layers at a few thousand degrees only.

Sirius, the brightest star in the sky, is linked with a companion which has a mass about four-fifths of that of the Sun, and gives little light. But in 1914 Adams saw from Mount Wilson that it was white-hot, so that its low emission must be due to small size. Its heavy mass and small size indicates the surprising density of a ton to the cubic inch—at the time an incredible result. But Einstein's theory requires that the spectral lines should be shifted towards the red by an amount proportional to the mass divided by the radius. Adams examined the spectrum and again got the same enormous density, and other dense stars have now been found.

With temperatures of millions of degrees, the maximum energy in a star is far above the visible spectrum and consists of X-rays and γ rays, which are very effective ionizing agencies. Inside a star then the atoms will be ionized, that is stripped of their outer electrons; this is so, even in ordinary stars, while in the very dense stars it is probable that the atoms consist of nuclei only, stripped bare of even the innermost ring of electrons.

Stellar Evolution. When classed in absolute magnitudes, there are stars of all values, but more in the highest and lowest groups than in the intermediate ones. Those in the two more populous grades—bright and faint—are called giants and dwarfs respectively. It was thought that they indicated the course of evolution of every star, which, beginning large and diffuse, shrinks with rising temperature, owing to gravitational contraction and radio-activity. It reaches a maximum output as a giant, and then cools, going through the same changes in reverse order with much less luminosity, which, since temperatures are the same, means smaller size; the star has become a dwarf—a small star, with little radiation. This scheme has been modified by bringing in the new physical ideas.

The probable age of stars was formerly estimated as five to ten millions of millions of years, but we shall see some reason to reduce this to some ten thousand millions. To provide for such ages as these, enormous supplies of radiant energy are needed, much more than gravitational contraction or radio-activity can supply. Einstein's theory suggested that the source might be found in the conversion of matter into radiation by methods we shall describe presently. Radiation exerts pressure, and therefore possesses momentum, by which the Sun loses mass at the rate of 360,000 million tons a day. It must have lost mass faster when larger and younger, and thus an upper limit can be found for its age, a limit of somewhere about eight million million years, agreeing well with the earlier independent estimate for the ages of the stars.

Kant and later Laplace tried to explain the origin of the solar system by imagining a nebula shrinking and revolving under its own gravitation. At various stages it was supposed to leave behind rings of matter which condensed into globular planets. But, as Chamberlin showed, for a mass of gas of the required dimensions, gravity would not overcome the diffusive effects of molecular velocities and radiation pressure, and Sir James Jeans has indicated that planetary condensations would not be formed.

But in the spiral nebulae we have bodies a million times larger than that imagined by Laplace, and on this scale gravitation will overcome both gas pressure and radiation pressure, and the nebula, instead of scattering, contracts and spins faster as Laplace supposed. The theory fails for the comparatively small solar system, but succeeds for a gigantic stellar galaxy.

Jeans has proved mathematically that a mass of rotating gas will form a double convex lens, the edge of which becomes unstable and forms two whirling arms. In them local condensations will occur, each of the appropriate size to form a star within the limits of size we know stars to possess. Spiral nebulae are the forerunners of stellar systems.

But our solar system, small enough for diffusive effects to take charge, gives another problem. A globule on the arm of a spiral nebula, if rotation is fast enough to cause disruption, is more likely, according to Jeans' mathematics, to form a double star with two partners waltzing round each other. But if, Jeans and Moulton and Chamberlin suggest, two smaller stars came near each other in the gaseous stage, tidal waves would be formed, and, if the stars approached within a certain critical distance, such a wave would shoot out a long arm of matter, which might break up into bodies like the Earth and the other planets. This would happen but rarely, and, even in the millions of galaxies, not many such systems are to be expected.

The spectra of spiral nebulae show most known lines displaced towards the red. By Doppler's principle this indicates retrocession. The velocity is greater in proportion to the distance, and the phenomenon is sometimes described as an expanding Universe.

Double stars show their structure by spectral lines oscillating in time with the revolution of the two stars. But, in some cases, the calcium or the sodium lines do not share in this periodic motion, but only move with our galaxy of stars. Calcium and sodium, and possibly other elements, therefore, are scattered through space, perhaps condensing in places into cosmic clouds. The density is in general extremely small, about 10^{-24}, one atom to a cubic centimetre. With this scattered tenuous matter, there are also the cosmic rays observed by Millikan and others, which, the evidence shows, come to us from outer space, either from the tenuous matter or from spiral nebulae.

It will be remembered that, on Einstein's relativity theory of gravitation, the presence of matter or an electro-magnetic field produces something analogous to curvature in the four-dimensional space-time continuum. Since there is matter scattered about the Universe, in the forms of stars, nebulae and cosmic clouds, we get this curvature, and if we regard time as flowing evenly onward, space itself must be curved and the Universe finite. Light emitted from one place and travelling forward long enough would return to its starting point. Einstein's estimate gives the radius of space as $9 \cdot 3 \times 10^{26}$ c.g.s. units, or about a thousand million light-years.

The final problem is to discover the source of the energy radiated by stars, including our own Sun, in whose light we live and move and have our being. The internal temperatures run to tens of millions of degrees, and the total output is too great to be supplied by gravitational contraction and the radio-activity of any terrestrial elements. Perhaps some more complex and vigorous radio-active elements exist in stars and emit more heat, but it seems that the conversion of matter into radiation is probably the chief source of the enormous output of energy.

Einstein's relation between energy and mass, $E = mc^2$, where c is the velocity of light, 3×10^{10} centimetres a second, shows that one gram is equivalent to 9×10^{20} ergs of energy. If we suppose a complete conversion, as by the mutual extinction of protons and electrons, the energy liberated would give the Sun an age of $1 \cdot 5 \times 10^{13}$ (15 million million) years. But, as there is no direct evidence for this theory, some prefer an alternative explanation.

The energy changes which accompany the transmutation of elements by bombardment in modern apparatus such as the cyclotron, and Aston's accurate measurement of atomic weights with the mass-spectrograph, show what large amounts of energy are involved in the conversion of hydrogen into other elements and give means of calculating them. For instance, the atomic weight of hydrogen is $1 \cdot 00813$ (oxygen $= 16$) and helium is $4 \cdot 00389$, so that four gram-atoms of hydrogen, or $4 \cdot 03252$, yield one gram-atom of helium and leave a mass of $0 \cdot 02863$ to be converted into radiation. This gives for mc^2 $0 \cdot 25767 \times 10^{20}$ ergs of energy. If we take one gram-atom of hydrogen instead of four, we should get $6 \cdot 4 \times 10^{18}$ ergs, which is equivalent to about 200,000 kilowatt hours. The energy is of course less than that of complete conversion, but if 10 per cent of the Sun's mass suffers transmutation from hydrogen into not-hydrogen, enough energy would be liberated to support the radiation for some ten thousand million years. This would be somewhat increased by the heat liberated by gravitational contraction and radio-activity. The stability which this theory gives to the Sun and stars is a great point in its favour.

The Beginning and End of the Universe. Kelvin's principle of the dissipation of energy, equivalent to Clausius' maximum entropy, was extended, from the isolated system for which it was proved, to cover the physical Universe. Whether this extrapolation was justified is another question. If the principle can be taken as applying on so much larger a scale, and if the Universe can fairly be treated as

THE STELLAR UNIVERSE

isolated, it follows that all temperature differences will continue to decrease, and the energy consequently become less and less available for the performance of useful work. This process will continue till all cosmic energy has become heat uniformly distributed at a constant temperature, and no further operations are possible in a dead Universe.

The new developments in physics and astronomy demand that fresh consideration should be given to this problem. However great be the output of energy of the stars, as long as it is finite, there must be not only an end but also a beginning. Jeans holds that 'everything points with overwhelming force to a definite event, or series of events, of creation at some time or times not infinitely remote'. The facts of artificial radio-activity, and the need for explaining the energy of the radiation from the Sun and stars, have forced us to believe that matter is continually passing into radiation. If this process goes on to the end, as Jeans says, 'there would be neither sunlight nor starlight, but only a cool glow of radiation uniformly diffused through space'. Thus, though the process is quite different, the outcome is the same as that deduced from the law of dissipation of energy—in the end no further action can occur and the Universe is dead.

Some people find the idea of the death of the physical Universe an intolerable thought. Perhaps to try to satisfy them, search has been made for natural means whereby the dissipation of energy or the destruction of matter might be reversed. Possibly there may be some physical process which plays the part of Maxwell's demon (p. 98), and separates out fast moving molecules. Again, there is another story which seems conceivable, though enormously unlikely, which runs as follows. If infinite time is available, all improbable things may happen. Chance concentrations of molecules might reverse the effect of random shuffling and undo the deadly work of the second law of thermodynamics; chance concentrations of radiant energy might saturate a part of space, and matter, perhaps one of our spiral nebulae, might crystallize out and start a new cycle. The probability against such a happening is fantastically great, but infinity is greater. However long it is necessary to wait for such a chance to occur, eternity is longer. Can we thus explain the course of past creation, and, when the present Universe has passed, apparently for ever, into 'a cool glow of radiation', may we imagine a new beginning? Probably not—probably something less unlikely would intervene this side of eternity.

It has been suggested that the second law of thermodynamics holds good only in an expanding Universe, and Tolman has formulated a scheme of relativistic thermodynamics in a contracting Universe in which the second law is reversed. This suggests the possibility of a pulsating Universe, in which we chance to be living in a phase of expansion and need not contemplate a beginning or an end. But these ideas are speculative, and the probabilities favour the dissipation of energy, and the transmutation of matter into a 'cool glow of radiation uniformly diffused through space'. There may be changes or reversals before or when this end is reached, but they are beyond the range of our present science.

Conclusion. And now we have to sum up the lessons we have learned and the outlook we have reached. Fifty years ago most men of science would have proclaimed a far more confident faith than their successors can preach to-day. Then it seemed that physics had laid down the main lines of inquiry once for all, and need only look to increased accuracy of measurement and a credible description of the luminiferous and electro-magnetic aether. Mach and Karl Pearson were teaching that science gave only a conceptual model of phenomena, but most men held a crude belief that physics revealed ultimate reality about matter and energy, and that natural selection fully explained organic evolution. Others went further and, carrying science into metaphysics, combined *Kraft und Stoff* with *Darwinismus* to formulate a philosophy of determinism and even of materialism.

To-day much of this nineteenth-century vision has vanished. Matter has ceased to be hard, impenetrable, material atoms, and become an amazingly complex structure, in which protons, electrons, positrons and neutrons jostle each other while waiting for new particles to be discovered. Radiation is no longer a train of regular mechanical waves in a semi-rigid medium or electro-magnetic waves in aether, but is now regarded as gushes or quanta of 'action', an incomprehensible product of energy and time. The heavier atoms

THE STELLAR UNIVERSE

are radio-active, and explode spontaneously, while some lighter ones are made radio-active by bombardment with α particles, and both are transmuted into other elements. All terrestrial atoms are open structures, in which empty space far transcends that occupied, though in dense stars the electrons are stripped off, leaving nuclei only—tons to the cubic inch.

Even with all this complexity, matter cannot be represented by particles only. The particles have to be imagined as accompanied by waves, or, to go farther, as themselves described by wave-mechanics. In a wave-group, there is an essential uncertainty in either the position or the velocity of a particle. The certainty of the older physics was part of the evidence on which philosophic determinism was built, and the principle of uncertainty destroys that part of the evidence, though of course it does not establish the opposite doctrine of human free-will.

The stellar Universe is far larger than was formerly realized. Our galaxy of stars is only one out of millions, the spiral nebulae being stellar galaxies in the making, some being so far away that their light takes hundreds of millions of years to reach us.

Thus the Universe is both larger and more complex than once appeared. We are not, as we thought, just about to understand it all. The larger the sphere of knowledge, the greater the area of contact with the unknown, and the farther we push into the unknown the less easy is it to represent what we find there in simple, understandable terms. Much of it must be left in mathematical equations. And, in any case, science can only build a model of nature, leaving to philosophy the problem of the possible reality which may or may not stand behind it.

The philosophy of Hegel revived the idea that a knowledge of the real world could be obtained *a priori* by logic. Such beliefs echo down the ages from Parmenides, Zeno and Plato. At the other extreme Francis Bacon exalted pure experiment, and held that, by an almost mechanical process, general laws could be established with certainty. But, in fact, scientific discovery involves both induction and deduction. First the right mental concepts must be picked out from the confused medley of phenomena. For instance, Aristotle's ideas of substance and qualities, natural places, etc. were useless as basic concepts for dynamics, and only led, if they led anywhere, to false conclusions. No advance was possible till Galileo and Newton, discarding the whole Aristotelian scheme, chose, as new fundamental ideas, length, time and mass, and thus were able to think in terms of matter and motion. This choice of concepts is itself a form of induction, and the most important step in framing scientific knowledge. Next, guesses at relations between the concepts, founded on preliminary observation, must be made. Their logical consequences must then be deduced by mathematics or otherwise, and these must be tested by further observation or experiment. An empirical rule, resting on facts alone, does not bring the conviction which follows an explanation of the law by an accepted theory. Even then, and more so with induction alone, the laws are only probably true, though the probability in favour of some of them may be so great as to approach, though never reach, certainty. A few years ago we all should have been willing to bet heavy odds that Newton's laws of gravity and the constancy of the chemical elements were accurately true, yet Einstein and Rutherford have proved us wrong.

Scientific philosophy has been much advanced by the application of mathematics. Lobatchevsky invented non-Euclidean geometry; Weierstrass proved that continuity does not involve infinitesimals; Cantor framed a theory of continuity and infinity which resolved Zeno's ancient paradoxes; Frege showed that arithmetic follows from logic; Russell and Whitehead traced fundamentals in their Principles of Mathematics. The new semi-realism, founded on these (with other) considerations, gives up the hope of explaining phenomena from a comprehensive theory of the Universe, as indeed science did when breaking with Scholasticism in the seventeenth century. Philosophy now fits its knowledge together piecemeal as does science, formulating hypotheses in the same way. But it goes beyond Mach's pure phenomenalism, and holds that science is concerned in some way with persisting realities. Nevertheless, as seen above, mental concepts are necessary for scientific analysis, and the relations which are called 'laws of nature' are relations between mental concepts and not between concrete realities. To suppose so is to fall into what Whitehead calls the 'Fallacy of Misplaced Concreteness'. The doctrine of mechanical determinism, he holds, only applies to abstract concepts; the concrete entities of the world are complete organisms, in which the structure of the whole influences the characters of the parts; mental states are part of the organism and enter into the whole and so into its parts.

In the seventeenth century two systems were simultaneously in being—Aristotelian and Newtonian. And a few years ago we had reverted to a similar dichotomy. The classical Newtonian dynamics still gave a useful method of solving many problems, though for others relativity and the quantum theory were necessary. As Sir William Bragg said, we freely used the classical theory on Mondays, Wednesdays and Fridays and the quantum theory on Tuesdays, Thursdays and Saturdays. But Eddington has now put together a new quantum synthesis, which points the way to a complete acceptance of the new views with the classical theory as a limiting case.

The older physics included a firm belief in cause and effect, a belief which began with the Greek atomists more than two thousand years ago. Even when the kinetic theory of gases dealt with the pressures due to molecular bombardment, the molecules were considered statistically in large numbers, and their joint effects were determined and calculable. But when C. T. R. Wilson and Rutherford began to trace individual atoms, prediction failed; there was no means of knowing which radio-active atom would explode next. Perhaps further investigation may reveal the finer structure of the atoms and bring each one into the realm of law and the region of prediction. But there is no sign of it as yet; all we can do is to calculate probabilities, and set forth the odds on one atom exploding in the next hour. Even if each atom became calculable as a whole, we cannot fix both the position and the velocity of its electrons; a fundamental principle of uncertainty seems to lie at the base of our model of nature.

The regularities of science may be put into it by our methods of observation or experiment. For instance, white light is an irregular disturbance into which regularity is put by our examination with prism or grating. Atoms can only be examined by external interference which must disturb their normal structure: Rutherford may have created the nucleus he thought he was discovering. From the latest point of view, substance vanishes, and we are left with form, in quantum theory with waves and in relativity with curvature.

NOTE ON BIBLIOGRAPHY

GENERAL HISTORIES OF SCIENCE

Many references to sources for this book will be found in *A History of Science and its Relations with Philosophy and Religion*, by Sir William Cecil Dampier (Cambridge, 1929, 1930 and 1942).

More detailed accounts within the limits of time indicated will be found in four books:

Science since 1500, by H. T. Pledge (London, 1939).

A Short History of Science to the Nineteenth Century, by Charles Singer (Oxford, 1941).

A History of Science, Technology and Philosophy in the Sixteenth, Seventeenth and Eighteenth Centuries, by A. Wolf (London, 1935, two volumes).

The most complete book of reference is *Introduction to the History of Science*, by G. Sarton (Washington, Vol. I, 1927; Vol. II, 1931, bringing it to 1300). Other volumes in preparation.

Cambridge Readings in the Literature of Science, by W. C. Dampier-Whetham and Margaret Dampier Whetham, now Mrs Bruce Anderson (Cambridge, 1928).

Current Literature will be found in the periodical *Isis* (Burlington, Vermont, U.S.A.).

MATHEMATICS

A Short Account of the History of Mathematics, by W. W. Rouse Ball (London, 1912).

A History of Mathematics, by F. Cajori (London, 1919).

The History of Mathematics in Europe, by J. W. N. Sullivan (Oxford, 1925).

History of Greek Mathematics, by Sir T. L. Heath (Oxford, 1921), shortened in his *Manual of Greek Mathematics* (Oxford, 1931).

ASTRONOMY

History of Greek Astronomy, by Sir T. L. Heath (Oxford, 1913).

History of Astronomy during the Nineteenth Century, by Agnes M. Clerke (London, 1902).

A Hundred Years of Astronomy, by R. L. Waterfield (London, 1938).

General Astronomy, by H. Spencer Jones (London, 1923).

The Birth and Death of the Sun, by G. Gamow (London, 1941).

PHYSICS

A Short History of Physics, by H. Buckley (London, 1927).

A History of Physics, by F. Cajori (London, 1929).

The Recent Development of Physical Science, by W. C. Dampier-Whetham (London, 5th ed. 1924).

Chemistry

A Short History of Chemistry, by J. R. Partington (London, 1937), and *Origins and Development of Applied Chemistry* (London, 1935).

A Hundred Years of Chemistry, by A. Findlay (London, 1937).

Biology and Medicine

A Short History of Biology and *A Short History of Medicine*, by C. Singer (Oxford, 1931 and 1928).

The History of Biological Theories, by E. Radl (Eng. Trans., Oxford, 1930).

The History of Biology, by E. Nordenskiold (London, 1929).

History of Botany, 1530–1860, by J. von Sachs (Oxford, 1906), and 1860–1900, by J. R. Green (Oxford, 1909).

Introduction to the History of Medicine, by F. H. Garrison (Philadelphia, 4th ed. 1929).

INDEX

Abelard, 41, 42
Aberration of light, 158
Ability, inheritance of, 124
Abram, 7
Absolute temperature, 98
Accademia Secretorum Naturae, etc., 66
Achaeans, 15
Achilles and the tortoise, 20
Acquired characters, 122
Actinium, 145
Action, 81, 82, 151, 152, 160, 170
Adam, N. K., 135
Adams, J. C., 69, 82
Adaptation, 118
Adonis, 141
Adrenalin, 136
Aeschylus, 16
Aether, 67, 68, 100, 103; motion of, 159
Agassiz, 119
Agrippa, Cornelius, 60
al-Batani, 39
Albertus Magnus, 43
al-Bīrūni, 39
Albury, 120
Alchemy, 31-2, 38, 48, 156
Alcmaeon, 19, 22, 41
Alexander, 12, 22, 24, 30
Alfred the Great, 37, 40
al-Ghazzali, 40
al-Haitham, 39, 44
al-Hazen, 39
al-Kindi, 39
Almagest, 39
Alpine race, 8
al-Rāzi, 39
Amber, 57
Amino-acids, 108, 135
Amontons, 93
Ampère, 102, 103
Anaemia, 138
Anaesthetics, 53
Analine, 106
Anatomy, 48
Anatomy, Physiology and Botany, 54-6

Anaxagoras, 17
Anaximander, 17
Ancient Learning, end of, 34
Anderson, Carl, 155
Andrews, 97
Animal spirits, 33, 54, 68
Animals, experiments on, 55, 85, 114, 116, 117, 136
Anselm, 41
Anthropoid apes, 123, 133
Anthropology, 123-5, 140-1
Antigens, 139
Antipyrene, 107
Aphrodite, 141
Apollonius, 31
Apple, Newton's, 68, 69
Appleton, Sir E. V., 158
Aquinas, St Thomas, 22, 40, 43, 45, 48, 50, 60
Arabia, 7, 12
Arabian School, 37-40
Arabic language, 37
Archaei, 84
Archimedes, 23, 27, 28, 47
Argon, 105
Aristarchus, 19, 27, 50
Aristotle, 17, 18, 21, 22-24, 26, 34, 37, 41, 42, 43, 45, 48, 50, 63, 74, 120, 138, 142, 171, 172, 173
Arithmetic, 9
Aromatic compounds, 106
Arrhenius, 110
Aryan people, 14
Ascorbic acid, 135
Aspirin, 107
Assyrians, 7
Astbury, 148
Aston, 18, 150
Astrology, 13, 25, 29, 31, 48, 67
Astronomy, 9, 13, 68-71, 81-3, 162-73
Asymmetric atoms, 107
Athens Academy, 34, 37
Atomic energy, 156
Atomic numbers, 149

INDEX

Atomic numbers and X-rays, 148
Atomic theory, 14, 18, 22, 25, 38, 104–5, 147, 148, 155, 165, 171
Atomic weights, 104
Atomists, 17–18, 80, 119, 173
Atoms, 76, 128
Atwater and Bryant, 115
Augustine, St, 35, 49
Australia, 89, 119
Averroes, 40
Avicenna (Ibn Sīnā), 39, 53
Avogadro, 104
Azores, 46

Babylonia, 12–14, 24
Bacon, Francis, 58, 120, 171
Bacon, Roger, 39, 42, 44, 45
Bacteria, 115, 117, 118
Bacteriology, 116–17
Baer, von, 114
Baghdad, 39, 40
Bahamas, 46
Bain, 127
Bakewell, 88
Baliani, 62, 70, 72
Banks, Sir Joseph, 89
Banting and Best, 136
Barnett, 158
Barometer, 65, 66
Barrow, Isaac, 64, 68
Barton, Catherine, 77
Base-line, 118
Basle, 53
Bateson, William, 129
Bauhin, J. and G., 56
Bayliss, 136
Beagle, 119
Beaumont, 114
Becquerel, Henri, 145
Bede of Jarrow, 37
Beginning and End of the Universe, 168–70
Behaviourism, 137
Bell, 115
Benzene, 106
Berengarius, 41
Beri-Beri, 134
Berkeley, Bishop, 78, 79, 81
Berkeley, Earl of, 110

Bernal, 148
Bernard, Claude, 114, 115
Bernouilli, James, 81, 95
Berossus, 29
Bertholet, 104
Beryllium, 154
Berzelius, 102, 106, 109
Bestiaries, 56
Bibliography, 174–5
Bills of Mortality, 123
Biochemistry, Physiology and Psychology, 134–8
Biology and its Effects, 113
Birth-rate, 132
Black, Joseph, 83, 93
Black Notley, 86
Blackett, 154
Blood corpuscles, 114, 115, 138
Boerhave, 83, 85
Boëthius, 37, 41
Bohr, Niels, 18, 142, 152, 153
Bologna, 40, 56
Boltzmann, 96
Boole, 125
Borch, 83
Botanic Gardens, 56
Botany, 33
Bothe, 154
Botticelli, 48
Boussard, 9
Boussingault, 114, 118
Boyle, Hon. Robert, 64, 83, 84, 94, 104
Boyle's Law, 97
Bradley, 83, 158
Bragg, Sir W. and Sir L., 148, 173
Brahe, Tycho, 52
Brain, 30, 85, 137
Brazil, 160
Britain, 30
Bronze Age, 2, 6
Brownian Movement, 111
Brünn, 129
Bruno, Giordano, 52
Brussells, 53
Buccaneers, 88
Büchner, 126
Buckland, 119
Buddha, 14
Buffon, 86

INDEX

Bunsen, 100
Byzantium (Constantinople), 37, 40, 42, 44, 47

Cabot, 89
Cagniard-de-Latour, 116
Calendar, 10, 13
Caloric theory, 94
Calorie, 115
Calorimeter, 94
Calvin, 49, 55
Cambridge, 40, 55, 70
Cambridge Platonists, 64
Cambridge University Library, 77
Cambyses, 12
Cannizzaro, 104, 105
Cantor, 172
Carbohydrates, 108, 115
Carbon, 105, 107
Carbon and nitrogen cycles, 117
Carbon dioxide, 83, 117, 118, 136
Carbon monoxide, 114
Carnot, Sadi, 97
Carnot's engine, 97, 110
Carotene, 134
Carpenter, 121
Catalysts, 109
Cathode Rays and Electrons, 143–5
Cattle, breeding of, 88
Cave man, 2, 3
Cavendish, Henry, 82, 84, 94, 101
Cayenne, 88
Cells, living, 114
Cellulose, 135, 149
Celsius, 93
Celsus, 33, 53
C.G.S. units, 93
Chadwick, 154
Challenger, 119
Chamberlin, 166
Chambers, Robert, 120
Champollion, 9
Charcot, 115
Charlemagne, 37, 40
Charles I, 55
Chemical action, 108–9, 112
Chemical affinity, 75, 108
Chemical elements, 64, 105
Chemistry, 38, 83, 84

Chemistry, industrial, 106, 107, 109
Chemistry and Medicine, 53–4
China, 7, 14
Chlorine, 83
Chlorophyll, 117, 118, 135
Christians, early, 34
Chromosomes, 130, 131
Chronometer, 90
Church, Fathers of, 34
Cicero, 28, 32, 54
Circulation of the blood, 33, 48, 55
Classification, 41
Clausius, 95, 97, 98, 102, 126, 168
Clay, 111
Clement IV, Pope, 44
Cleopatra, 12, 29
Clifford, 125
Cloud nuclei, 144, 147
Cnidos, 20
Cnossus, 14, 15
Coaches, stage, 91
Coagulation, 111, 112, 134
Coal-tar, 106
Cockcroft, 156
Colbert, 88
Colchester, 56
Colloids, 111–12, 116, 139
Columbus, 30, 46
Combining volumes of gases, 104
Combining weights, 104
Combustion, 106
Composition, percentage, 106
Compton, 151
Conclusion, 170–3
Concreteness, fallacy of, 172
Conductivity of gases, 145
Conductivity of liquids, 109
Conductors and insulators, 101
Conduitt, John, 77
Consciousness, 128, 137
Conservation of Energy, 95, 115, 126
Constantinople, 85; *and see* Byzantium
Cook, James, 89
Co-ordinate geometry, 58
Copenhagen, 52, 152
Copernicus, 19, 43, 49–53, 61, 62
Cordus, Valerius, 56
Cork, Earl of, 65
Corpuscles, 144

INDEX

Correlation of forces, 94
Cos, 20
Cosmic clouds, 167
Cosmic physics, 101
Cosmic rays, 154, 155, 167
Cotyledons, 87
Coulomb, 101
Creation, 122, 169, 170
Crete, 12, 14–15
Crew, 131
Crommelin, 160
Crookes, Sir William, 146, 147
Crystalloids, 111
Crystals, 107
Curie, M. et Mme, 145, 146
Curtius, 108
Curvature in space-time, 167
Cuvier, 90
Cyclotron, 156
Cyril the Patriarch, 36
Cyrus, 12

Dalton, John, 18, 104, 105, 129, 142
Dampier, W. C. D., 110, 112
Dampier, William, 88, 89, 119
d'Anville, 119
Dark Ages, 36–7
Darwin, Charles Robert, 113, 119, 120, 121, 122, 123, 126, 127, 132
Darwin, Erasmus, 121
Darwin, R. W., 121
Darwinismus, 126
Davisson, 152
Davy, Sir Humphry, 102, 117
Dawson, 133
Death of the Universe, 169
de Broglie, 153
Debye, 151
Decay, rate of, 146
Decimal system, 9, 13
Decimal units, 93
Declination, diagram of, 57
Deflexion, electric and magnetic, 143, 149
Defoe, 89
Degrees of freedom, 97
Dehydrogenases, 136
Democritus, 17, 18, 22, 23, 25, 104
Demon, Maxwell's, 98, 99, 169

Density, 23, 27, 71
Descartes, René, 58, 59, 67, 68, 74, 120, 128
Determinism, 137, 170, 172
Deuterium, 150
Deuteron, 150
de Vilmorin, 131
de Vries, 129
Dewar, Sir James, 97
Dextrose, 114
Diana Nemorensis, 4
Dielectric constant, 103
Differential equations, 81
Diffraction, 148, 153
Digby, Sir Kenelm, 64
Digges, Thomas, 52
Diogenes the Babylonian, 32
Dionysos, 141
Diophantus, 33
Dioscorides, 33, 56
Disraeli, 122
Dissipation of energy, 168, 169
Distemper, 139
Dominance, 129
Donnan, F. G., 135
Doppler, 101, 167
Dorians, 15
Dreams, 6
Drosophila, 130
Dualism, 59
Dubois (Sylvius), 54, 123
Dumas, 114
Dunkin, 139
Duns Scotus, 45, 58
Dürer, Albrecht, 48
Dynamics, 61–3
Dynamo, 103

Earth, size of, 30
Earth solidification, 1
Earthquakes, 133
Easter, date of, 41
Eclipses, 13, 20
Ecology, 132
Ecphantus, 30
Eddington, Sir Arthur, 160, 161, 164, 173
Eels, life of, 134
Effervescence, 84

INDEX

Egypt, 8–12, 24, 133, 140
Ehrlich, Paul, 107, 139
Eighteenth Century, 78–91
Einstein, Albert, 26, 82, 151, 159, 160, 161, 165, 166, 168, 172
ἡ κοινή, 24
Electric charge, 102, 111, 112
Electric Current, 101–3
Electric units, 101
Electrical engineering, 103
Electro-chemical equivalent, 102
Electro-chemistry, 102
Electrolysis, 102, 109, 110
Electro-magnetic waves, 103, 144, 148, 157–8
Electro-magnetism, 103–4
Electromotive force, 103
ἤλεκτρον, 57
Electron microscope, 138, 158
Electrons, 128, 147, 148
Electron-volts, 155
Elements, 17
Eleusinian mysteries, 16
Embalming, 11
Embryology, 114
Empedocles, 17
Encyclopédie, 79
Endeavour, 89
Endocrinology, 136
Energy, 95, 160, 167, 168; kinetic, 66, 95, 96
Engels, 126
Engine, theory of, 97, 98, 110
England, 37
Entropy, 98, 126, 160, 168
Enzymes, 109, 116, 136
Eoliths, 2
Ephemerides, 49
Epicurus, 17, 25, 26
Equilibrium, physical and chemical, 99, 108
Equinoxes, 29
Erasistratus, 30
Erasmus, Desiderius, 49
Eratosthenes, 30
Ergosterol, 134
Erlanger, 137
Error, curve of, 96, 124
Ether, 53

Etruria, 121
Euclid, 26, 27, 29, 39, 41, 172
Eudemus, 30
Euler, 81, 134
Euphrates, 7
Europe, peoples of, 8, 15, 140
Evans, Sir Arthur, 14
Evolution and Natural Selection, 113, 119–23
Expanding Universe, 167

Fahrenheit, 93
Faraday, Michael, 92, 102, 103, 109, 112
Fathers of the Church, 34–6
Fats, 108, 114, 115
Feather, N., 154
Ferdinand and Isabella, 46
Ferments, 55, 108, 114, 116
Fermi, 154
Fertilization, 131
Fire, 2, 5, 83
Fischer, Emil, 106, 108, 135
Fischer, O., 107
Fish, migration of, 133
Fisher, R. A., 132
Fitzgerald, G. F., 159
Fizeau, 158
Fletcher, Sir W. M., 136
Flint implements, 2, 119
Florence, 66
Fluxions, method of, 70, 71, 81
Foot and mouth disease, 138, 139
Formulae, empirical and constitutional, 106
Fossils, 48, 119
Foucault, 100, 159
Fourneau, 139
Fourth dimension, 160
Fowler, Sir R. H., 164
Franck, 152
Franklin, Benjamin, 101
Fraunhofer, 100
Frazer, Sir James, 4, 140, 141
Free will, 171
Freedom, degrees of, 97, 99
Freezing point, 110
Frege, 172
French Guiana, 88

INDEX

Fresnel, 99
Freud, 137
Friedrich, 148
Fructose, 106
Fruits, genetics of, 130

Galen, 33, 36, 53, 55, 68
Galileo, 22, 24, 43, 45, 51, 61–4, 68, 70, 72, 73, 75, 84, 93, 100, 129, 172
Galileo and Newton, 61–77
Gall, 115,
Galton, Sir Francis, 124
Galvani, 101
Gas, 53
Gases, kinetic theory of, 66, 95, 96, 97
Gaskell, 137
Gassendi, 67
Gasser, 137
Gauss, 96, 101
Gay-Lussac, 98, 104
Geber, 38
Genes, 130, 131
Genetics, 129–33
Geneva, 55
Geography and Geology, 88–90, 118, 119
Geological periods, 1
Geology and Oceanography, 88, 89, 133–4
Geometry, deductive, 26
Geometry, non-Euclidean, 172
Gerard, John, 56
Germ cells, 122, 129, 130
Germer, 152
Gibbon, 35
Gibbs, Willard, 99, 111
Gilbert, William, 56, 57, 62, 67, 71
Gilbert and Lawes, 118
Glaisher, J. W. L., 69
Glands, 116
Glucose, 106
Glycerine, 114
Glycogen, 114, 136
Gnomon, 17
Goethe, 125
Golden Bough, 4, 140
Gonville and Caius College, 55
Graham, Thomas, 111, 112
Gratings, optical, 101, 148, 149

Graunt, John, 123
Gravity, 69, 82
Gray, Asa, 121
Greece and Rome, 16–33
Greek medicine, 20
Greek religion, 16
Green, 99
Gregory XIII, Pope, 11
Gregory of Tours, 36
Grew, N., 86
Grey, Edward (Viscount), 88
Grimaldi, 73
Grosseteste, 42, 44
Grotthus, 102
Grove, Sir W. R., 94
Guettard, 90
Guinea, Gulf of, 160
Guldberg and Waage, 109
Gulliver's Travels, 89

Haeckel, 126
Haematin, 135
Haemoglobin, 114, 135
Haemophilia, 135
Hales, Stephen, 83, 85
Haller, A. von, 85
Halley, 70
Hanno, 30
Haploid number, 130
Hardy, Sir W. B., 111, 136
Harkins, 154
Harun-al-Rashid, 37
Harrison, John, 90
Hartley, E. G. J., 110
Harvard observatory, 164
Harvey, William, 33, 48, 55, 67, 114
Head, Sir Henry, 137
Heat and Energy, 64, 66, 74, 93–7
Heath, Sir Thos., 28
Heaviside, 158
Hecateus, 20
Hegel, 120, 125, 171
Heidelberger, 139
Heisenberg, 153
Helium, 97, 100, 105, 152
Hellenistic Period, 24–6
Hellriegel and Wilfarth, 118
Helmholtz, H. von, 102, 127
Helmont, van, 53, 54, 55, 84

INDEX

Henson, 119
Heraclitus, 18, 19
Herbals, 56
Heresy, 35, 60
Hermes, 41
Hero, 31
Herodotus, 20
Herophilus, 29, 30
Herschel, 100, 163
Hertz, 103, 152, 157
Herzog, 148
Hicetas, 50
Hiero, 27, 28
Hieroglyphs, 9
Hilfield, 6
Hill, A. V., 137
Hindu numerals, 39
Hipparchus, 21, 27, 28, 29, 49, 163
Hippocrates, 20, 22, 36, 53
History, Dawn of, 7
Hittorf, 109
Hobbes, Thomas, 59, 64, 68, 80
Hofmann, 106
Hohenheim, T. von (Paracelsus), 53
Homberg, 83
Homer and Hesiod, 16
Hooke, 74, 84, 99
Hooker, Sir Joseph, 119, 121
Hopkins, Sir F. G., 134, 136
Hormones, 109, 136
Horsley, Sir Victor, 137
Horus, 9
Hounslow Heath, 118
Humboldt, von, 119
Hume, David, 79, 81
Hutton, James, 48, 90
Huxley, T. H., 119, 121, 123
Huygens, 64, 66, 67, 69, 72, 73, 74
Hybridization, 88
Hydrodynamics, 47
Hydrogen, atomic weight of, 167; liquefied, 97
Hydrogen spectrum, 152, 153
Hydrophobia, 116
Hydrostatics, 47
Hypatia, 36

Iamblichus, 36
I-am-hotep, 11

Iatrochemists (Spagyriṣts), 53, 65
Ibn-Junis (Yūnus), 39
Ibn-Sīnā, 39, 53
Ice Age, 3, 119
Immanence, 65
Immunity, 139
India, 14, 30
Indigo, 107
Indus, 7
Influenza, 140
Inheritance, 120, 131, 132
Innocent VIII, Pope, 60
Inoculation, 85, 139
Inquisition, 52
Interference, 99, 148, 152, 153
Introspection, 127
Inverse Square Law, 69, 70, 101
Inversion of sugar, 108
Ionian philosophers, 16–17
Ionosphere, 158
Ions, 102, 109, 110, 142, 144
Ireland, 37
Iron age, 7, 14
Ishtar, 141
Island universes, 164, 166
Isobars, 150
Iso-electric point, 111
Isomerism, 106
Isotopes, 150
Ivanovski, 117, 138

Jabir-ibn-Haiyan, 38
Jamaica, 88
James II, 77
Java, 123
Jeans, 166
Jenner, Edward, 85, 116
Jerome, 36
Jesty, Benjamin, 85
Jevons, 125
John of London, 44
Johnson, Dr, 79
Joule, J. P., 94, 95, 96, 98
Jundishapur, 37
Jupiter's satellites, 61, 158
Justinian, 37

Kamerlingh Onnes, 97
Kant, 79, 82, 120, 125, 166

Karnak, 11
Kaufmann, 143, 160
Kekulé, 106
Kendal, 139
Kennelly-Heaviside layer, 158
Kepler, John, 52, 62, 67, 70, 82, 162
Keratin, 148
Keynes (Lord), 77
Kilogram, 93
Kinetic theory, 66, 95, 96, 97, 147
Kipping, 148
Kirchhoff, 100
Koch, 117
Kolrausch, 109
Krebs, 136
Krypton, 105
Kühne, 116
Kunsman, 152

Laborde, 146
Lagrange, 80, 81, 82
Laidlaw, Sir P., 139
Lamarck, 120, 122
Landed families, 132
Langley, 137
Langmuir, 135
Language, 78
Laplace, P. S. (Marquis de), 80, 82, 96, 166
Latent heat, 94, 99
Latitude, 89
Laue, 148
Lavoisier, A. L., 84, 95, 104, 106, 125
Lawes and Gilbert, 118
Lawrence, A. E., 156
Laws of Nature, 58, 75, 172
Lead, 146
Le Bel, 107
Leguminous plants, 118
Leibniz, 79, 81, 120
Lenard, 152
Leo X, Pope, 49
Leo XIII, Pope, 45
Leonardo da Vinci, 28, 47–8, 54, 90
Leucippus, 17
Lever, theory of, 27, 47
Leverrier, 82
Lewis, 150
Leyden, 56, 83

Library of Alexandria, 30, 36
Liebig, 106, 107, 117, 118
Life, basis of, 84, 115, 128, 138
Life on other worlds, 162
Light, theory of, 99–100, 153; velocity of, 74, 83, 103, 158, 159
Limericks, 79
Lindemann (Lord Cherwell), 151
Linder and Picton, 112
Linnaeus (Carl von Linné), 87
Liquefaction, 97
Lisbon, 56
Lister (Lord), 117
Lobatchevsky, 172
Local motion, 62
Locke, John, 78, 81, 125, 127
Lodge, Sir Oliver, 110, 159
Löffler and Frosch, 117
Longitude, 89
Loom, 12
Lorentz, 144, 145
Lot, 7
Louis XIV, 88
Lower, 84
Lubbock (Lord Avebury), 121
Lucretius, 17, 18, 25, 32
Lunar theory, 89, 90
Luther, Martin, 49
Lyell, Sir Charles, 119, 121, 123
Lymington (Viscount), 77

MacCullagh, 99
Mach, E., 71, 72, 113, 125, 170, 172
Machinery, 90–1
Maclaurin, 81
Macrocosm, 19, 20, 21, 25, 42
Madras, 88
Magic, 4, 5, 16, 25, 35, 48, 56, 60, 87, 141
Magnetic compass, 46, 57
Magnetic permeability, 103
Magnetic storms, 119
Magnetism and Electricity, 56–7
Maimonides, 40
Maintenance ration, 115
Majendie, 115
Malaria, 36, 117
Malebranche, 64
Malpighi, 56, 86

Maltese fever, 117
Malthus, Robert, 120, 121
Mammoths, 3, 4
Man, age of, 1, 119
Manichaeism, 35, 60
Maps, 30, 119
Marcus Aurelius, 32
Mariotte, 64
Marsh, Adam, 44
Marshall Hall, 115
Marx, 126
Mass, 47, 57, 62, 63, 71, 160
Mass law, 109
Materialism, 80, 81, 125, 126, 170
Materials, strength of, 63, 91
Mathematics, 48
Mathematics and Astronomy, 81-3
Maupertuis, 81
Maury, 119
Maxwell, James Clerk, 92, 96, 98, 99, 100, 101, 103, 104, 144, 157, 160, 169
Mayer, J. R., 94
Mayo, 84
Medicine, 11, 20, 33, 36, 38
Mediterranean race, 8, 15, 140
Mela, Pomponius, 32
Meldrum, 136
Melvill, 100
Memphis, 9
Menageries, 86
Mendel, Abbot G. J., 123, 129, 130
Mendeléeff, 105
Mendelians and Biometricians, 132
Menes, 9
Mental power, 124
Mercuric oxide, 83
Mercury, orbit of, 161
Merton College, 55
Mesopotamia, 133
Mesotrons, 155
Meteorology, 119
Metre, 93
Mexico, 46
Michell, 82
Michelson and Morley, 159
Microcosm, 19, 20, 21, 25, 42
Microscope, 86; electron, 138, 158; ultra, 111
Middle Ages, 34-45

Migration of fish, 133
Milky Way, 154, 163
Mill, J. S., 127
Millikan, 155, 167
Millington, Sir Thomas, 86
Miners' cramp, 138
Minkowski, 160
Minos, 15
Minot, 138
Mint, 77
Mithras, 35
Mitscherlich, 107
Mohl, 115
Molecules, 104
Mondino, 54
Monomolecular change, 147
Montagu, Lady Mary Wortley, 85
Moon, 61, 69, 89
Morgan, Lloyd, 137
Morgan, T. H., 130
Morton, 117
Moseley, H. G. J., 149
Moulton, 166
Mount Wilson Observatory, 163, 165
Mousterian civilization, 3
Muhammad, 37
Müller, Johan, 49, 113
Museum of Alexandria, 29, 30, 36
Mutations, 129
Myosin, 135
Mystery religions, 25, 35, 141
Mysticism, 45
Mythology, 16, 140, 141

Naples, 66
Napoleon, 80
Natural History of Selborne, 88
Natural philosophy, 66, 70
Natural selection, 120-3
Navigation, 46, 88, 89, 90, 119
Neanderthal man, 3, 123, 133
Nebulae, spiral, 82, 164, 167
Nebular theory, 82, 166
Neddermeyer, 155
Nemi, 4
Neolithic man, 2, 5, 133
Neon, 105, 150
Neo-Platonism, 34, 35, 37, 41, 42, 50
Neptune, 82

Nernst, 151
Nervous system, 85, 114, 115, 116, 128, 134, 136, 137
Nestorius, 37
Neutrons, 154
Newcomen, 90
New Holland (Australia), 89, 119
New Physics and Chemistry, 141–61
Newton, Sir Isaac, 21, 22, 24, 29, 59, 62, 63, 64, 66, 68–77, 81, 82, 84, 87, 89, 94, 95, 100, 104, 108, 125, 129, 142, 152, 153, 160, 161, 162, 172, 173
Nicholas of Cusa, 45, 58
Nicholson and Carlisle, 102
Night blindness, 134
Nile, 7, 8, 10
Nineteenth-Century Biology, 113–28
Nineteenth-century Science and Philosophy, 124–7
Nitrogen cycle, 118
Nomads, 7
Nominalism, 24, 41, 42, 45
Nordic race, 8, 15, 140
Novara, 50
Number, 18, 19, 50, 52
Nürnberg, 49

Occam, William of, 45, 58
Oceanography, 89, 119, 133
Ochus, 12
Oersted, 102
Ohm, G. S., 103
Ohm's Law, 103, 109
Old Testament, 7, 8
Omar Khayyám, 39
Optics and Light, 73–5
Organic Chemistry, 105–8
Organism, 113, 128, 138, 173
Origen, 35
Origin of Species, 121, 122
Origins, 1
Orphic mysteries, 16
Orthodoxy, 35
Oscillation, centre of, 66
Osiander, 50
Osmotic pressure, 110, 135
Owen, Sir Richard, 122
Oxford, 40, 55
Oxygen, 83, 84

Padua, 56
Palaeoliths, 2, 3
Palos, 46
Palstave, 6
Pancreas, 114
Paracelsus (*see* Hohenheim), 53
Parallax of stars, 163
Paris, 40, 43, 67
Parliament, 77
Parmenides, 171
Parthenogenesis, 131
Particles, list of, 155
Pascal, Blaise, 65, 66, 96
Pasteur, Louis, 107, 116, 117
Paul, St, 35
Pavlov, 137
Pearson, Karl, 131, 170
Peas, green, 129
Pendulum, 62, 66
Peptones, 108
Periodic table, 105, 149
Perkin, W. H., 106
Pernicious anaemia, 138
Perrault, 90
Persia, 7
Peru, 46
Peter de Maharn-Curia, 44
Petrarch, 46
Petty, Sir William, 123
Pfeffer, 110
Pharmacy, 33
Phase Rule, 99
Phenacetin, 107
Phenomenalism, 172
Phenylglycine, 107
Philae, 12
Philip of Macedon, 22
Philosophy, 58–9, 78–81, 172
Phlogiston, 83, 84
Photons, 151, 155
Physical constants, 161
Physics and Chemistry of the Nineteenth Century, 92–112
Physiological apparatus, 113
Physiology, 30, 48, 113, 116, 134–8
Physiology, Zoology and Botany, 84–8
Picard, 69
Pictures in caves, 2, 3, 4
Pilgrim Trust, 77

INDEX

Pillars of Hercules, 30
Piltdown, 133
Pisa, 56
Pitchblende, 145
Pithecanthropus Pekinensis, 133
Planck, 142, 151, 153
Planets, 162, 167
Plankton, 119, 134
Plants, classification of, 87
Plants and Animals, 114, 117, 118
Plato, 16, 21, 22, 23, 26, 31, 37, 41, 42, 171
Platonism, 34, 35, 52
Pliny, 32, 37, 56
Plutarch, 28, 32, 50
Pluto, 162, 163
Polarization, electric, 109
Polarization of light, 99, 107
Polonium, 145
Polyneuritis, 134
Polypeptides, 135
Polyploidy, 130
Porphyry, 36, 41
Portsmouth, Earl of, 77
Portuguese exploration, 46
Poseidonius, 30, 32, 46
Positive electricity, 145
Positive Rays and Isotopes, 149–50
Positrons, 155
Potassium, 102
Potential thermodynamic, 99
Pottery, 8, 14
Prehistory, 1–7
Pressure of light, 101
Prices, 36, 46
Priestley, Joseph, 83, 84, 101
Primary and Secondary qualities, 63, 65
Principia, 70
Principles, primary, 38, 39, 65
Printing, 47
Probability, theory of, 66, 96, 112, 129, 173
Proclus, 37
Proteins, 108, 114, 115, 135, 148
Protons, 147, 148
Protoplasm, 111, 115
Prout, 105, 150
Proxima centauri, 163
Psychology, 72, 78, 127, 128

Psycho-physical parallelism, 128
Ptolemy I, 29
Ptolemy the Astronomer, 21, 29, 30, 39, 49
Ptolemy Epiphanes, 9
Putrifaction, 116
Puy-de-Dôme, 66
Pyramids, 12
Pythagoras, 18, 19, 26
Pythagoreans, 18–20, 50
Pytheas, 30

Quadrivium, 37, 41
Quantum mechanics, 153
Quantum theory, 142, 151, 153
Quetelet, 123, 131

Rabies, 116
Racemates, 107
Radiation, 142, 151, 169, 170
Radicals, 107
Radio-activity, 145–8
Radio-location, 158
Radio-speech, 157
Radium, 145, 156
Rainbow, 73
Rankin, 95
Rattlesnake, 119
Raven, C. E., 86
Ray, John, 83, 85, 86, 87, 88, 90
Rays, α, β, γ, 146
Realism, 23, 41, 42, 45
Reality, 78, 79, 170, 171
Réaumur, 93
Recent Biology, 129–41
Recessive characters, 129
Redi, 86
Reflex action, 115
Reflexes, 137
Reformation, 49
Regularities of science, 173
Relativity, 158–61
Renaissance, 46–60
Resistance, electric, 103
Resonance, 100
Reversible processes, 160
Reversible reactions, 108, 109
Richardson, Sir O. W., 157
Richer, 83, 88

INDEX

Rickets, 134
Rings, Newton's, 73
Ritual, 16, 140
Rivers, 7
Rivers, W. H., 4, 140, 141
Robinson Crusoe, 89
Roebuck, 89
Roemer, 74, 158
Roman Age, 32-3
Rome, 33, 36, 49, 66, 164
Röntgen, 142, 145
Rosetta stone, 9
Ross, Sir James, 119
Rothamsted, 118
Roughton, 136
Royal Institution, 102
Royal Society, 66, 67, 77, 89
Rumford, Count, 94
Russell, Bertrand (Earl), 172
Rutherford, Lord, 18, 142, 145, 146, 147, 154, 157, 172, 173

Saint-Hilaire, 120
Salerno, 36, 40
Salvarsan, 107, 139
Sanctorius, 54
Sargasso Sea, 134
Saturn's rings, 66
Sceptical Chymist, 65
Scheele, 83
Schiff, 116
Schleiden, 114
Scholasticism, 22, 34, 40, 42-5, 49, 50, 58, 67, 172
Schrödinger, 153
Schwann, 114, 116
Scientific Academies, 66-8
Scientific Age, 92-3
Scot, Reginald, 60
Scotland, 37
Scurvy, 89, 134
Secchi, Father, 164
Second, 93
Seleucus, 28
Semi-realism, 172
Semites, 7
Seneca, 32
Sensation, 114, 127
Sepia, 23

Servetus, 55
Sex determination, 131
Sheep, breeding of, 88
Sherrington, Sir Charles, 131, 136, 137
Signatures, doctrine of, 56
Silkworm disease, 116, 138
Simpson, 117
Sirius, 165
Smallpox, 85, 135
Smeaton, 91
Smith, Kenneth, 139
Smith, William, 90
Snell, 73
Socrates and Plato, 20-22
Soddy, 147, 149
Sodium, 102
Soil, structure of, 111, 118
Solander, 89
Solar system, 88, 162, 166
Solution, 109-11
Sophocles, 16
Sound, 82, 127
Space, 26, 38, 76, 142, 159, 160, 167
Spagyrists, 53, 65
Spain, 39, 42
Spallanzani, Abbé, 86
Specific gravity, 23, 27, 47
Specific heat, 94, 151
Spectrum analysis, 100-1
Spee, Father, 60
Speech, 127
Spencer, Herbert, 120, 125
Spinoza, 64
Spirits of the Wild, 5
Spiritus nitro-aereus, 84
Spontaneous generation, 86
Spy, 123
Stahl, 83, 84
Standards of length, etc., 13
Stanley, W. M., 138
Starch, 114
Starling, 136
Stars, age of, 165; cepheid, 163; classification of, 164; double, 163, 167; energy of, 165, 166; giant and dwarf, 165; heavy, 165; magnitudes of, 163; and nebulae, 162-4; number of, 163; spectra of, 164; streams of, 163; temperatures of, 164; variable, 163

Statistics, 96, 123, 173
Steam engine, 90
Steam ships, 91
Stellar evolution, 165, 168
Stellar Universe, 162–73
Stellar velocities, 101
Stevinus (Stevin), 61, 62, 72
St John's College, Cambridge, 56
Stoicism, 25, 32
Stokes, Sir G. G., 100
Stonehenge, 140
Stoney, Johnston, 145
Structure of atom, 150–4
Structure of stars, 164–5
Sugar, 135
Sulphaguanidine, 139
Sulphanilamide, 139
Sumatra, 88
Sumer, 12
Sun, age of, 166, 167; distance of, 162; energy of, 117, 118, 167, 169
Sun and moon, 28, 29, 50
Survival of the fittest, 127, 132
Sutton, 130
Swift, 89
Sydenham, 78
Sylvius (Dubois), 54, 55, 83, 84
Synthetic drugs, 139
Syracuse, 28
Syria, 8, 12
Syriac language, 37

Tahiti, 89
Tammuz, 141
Taoism, 14
Tartaric acid, 107
Taylor, 81
Telescope, 61
Temperature, 93, 98
Temperatures of Sun and stars, 101
Tertullian, 41
Thales, 16, 26
Theophilus, 30, 36
Thermionic valve, 157
Thermodynamics, 95, 97–9, 101, 110, 111; Second Law of, 169, 170
Thermometer, 93
Thompson, Benjamin (Count Rumford), 94

Thomson, Sir G. P., 152
Thomson, Sir J. J., 18, 142, 143, 144, 145, 146, 147, 149
Thomson, William (Lord Kelvin), 95, 97, 98, 126, 168
Thoth, 9, 41
Thucydides, 20
Thyroid gland, 116
Tiberius, 33
Tides, 32, 70
Tigris, 7
Time, 38, 76, 142, 159, 160
Toleration, 34, 78
Tolman, 170
Tombaugh, 162
Torricelli, 66
Torsion balance, 82
Torture, 60
Townsend, Sir J., 144
Transmutation of elements, 154–7
Travel literature, 89
Trigonometrical Surveys, 118–19
Trinity College, Cambridge, 68, 74, 77, 86
Triphenyl methane, 107
Trivium, 41
Turner, William, 56
Twins, 131
Tylor, 141
Types, theory of, 107

Ultra-microscope, 111
Uncertainty, principle of, 154, 171, 173
Uniformitarian theory, 48, 90, 119
Uniformity, Act of, 86
Universals, 41
Universe, contracting, 170; diagram of, 51; energy of, 95, 98; expanding, 167; limits of, 167
Universities, 40, 66
Unverdorben, 106
Ur of the Chaldees, 12
Uranium, 145, 146
Uranium X, 146
Uranus, 82
Urea, 106

Vaccination, 85, 116
Valency, 105, 112

INDEX

Van der Waals, 97
Van't Hoff, 107, 109, 110
Variation, calculus of, 81
Varro, 32
Velocity of escape, 162
Venel, 83
Venus, transit of, 89, 162
Vernalization, 132
Vesalius, Andreas, 54
Viète, François, 53
Vinogradsky, 118
Virchow, 115
Virgil, 32, 118
Viruses, 117
Viruses and Immunity, 138–40
Vitalism, 84, 115, 128, 138
Vitamins, 134, 135
Volta, 101
Voltaire, 80, 81

Wallace, A. R., 120, 121
Wallop family, 77
Walton, 156
Washburn, 150
Water, 53, 83, 84, 102, 104; heavy, 150
Waterston, 95
Watson, J. B., 137
Watt, James, 91
Wave-length of light, 99
Wave mechanics, 152
Wave theory of light, 99–100
Waves, 47, 48, 82, 103
Weber, E. H. and E. F., 116, 127
Wedgwood, Josiah, 121
Weierstrass, 172
Weight, 71, 72

Weismann, 122
Weyer, John, 60
Wheat, 13, 118
Whewell, 102
White, Gilbert, 88
Whitehead, A. N., 45, 172
Wiechert, 143
Wieland, 136
Wien, 149
Wilhelmy, 108
Williamson, A. W., 109
Willstätter, 135
Willughby, Francis, 86
Wilson, C. T. R., 144, 147, 173
Wislicenus, 107
Witchcraft, 60
Wöhler, 106
Wolff, 114
Wollaston, 100, 101
Wood, T. B., 115
Woodward, John, 90, 133
Woolsthorpe, 68, 69
Wundt, 127
Würzburg, 60

Xenon, 105
Xenophanes, 16
Xenophon, 21
X-rays, 135, 142, 143, 145, 149
X-rays and Atomic Numbers, 148–9

Young, Thomas, 9, 95, 99

Zeeman, 145
Zeno of Citium, 25, 32
Zeno of Elea, 20, 171, 172

www.ingramcontent.com/pod-product-compliance
Ingram Content Group UK Ltd.
Pitfield, Milton Keynes, MK11 3LW, UK
UKHW040656180125
453697UK00010B/204